Original
FARMALL
Hundred Series
1954–1958

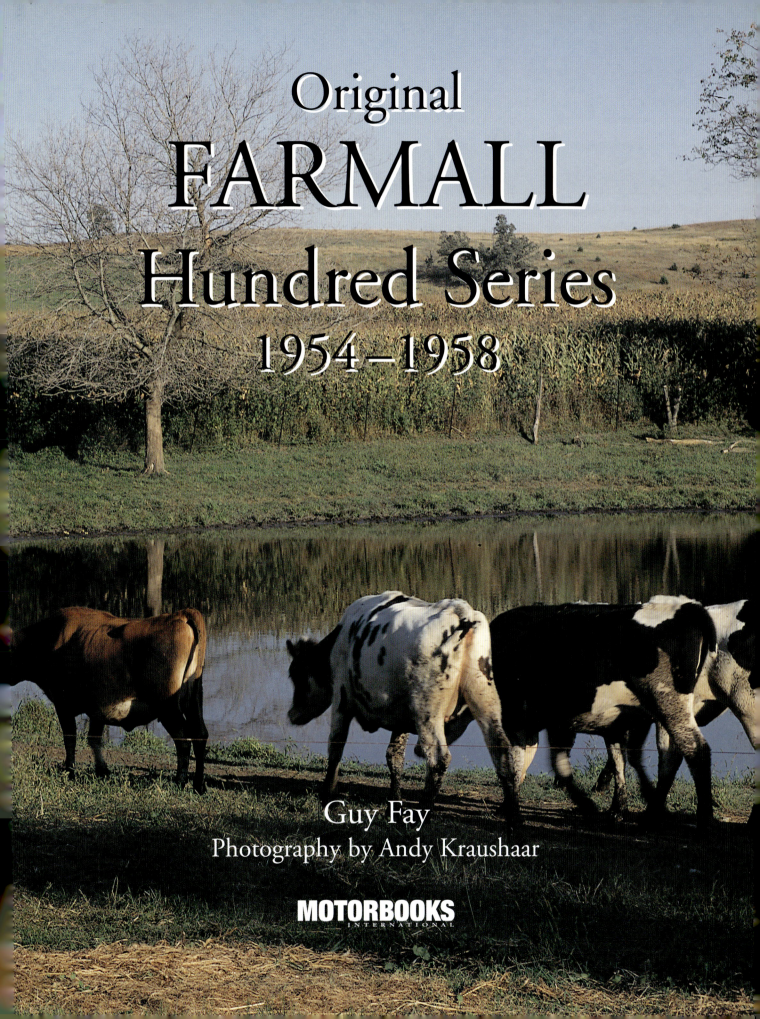

First published in 2003 by Motorbooks International, an imprint of MBI Publishing Company, Galtier Plaza, Suite 200, 380 Jackson Street, St. Paul, MN 55101-3885 USA.

© Guy Fay, 2003

All rights reserved. With the exception of quoting brief passages for the purposes of review, no part of this publication may be reproduced without prior written permission from the Publisher.

The information in this book is true and complete to the best of our knowledge. All recommendations are made without any guarantee on the part of the author or Publisher,
who also disclaim any liability incurred in connection with the use of this data or specific details.

We recognize that some words, model names and designations, for example, mentioned herein are the property of the trademark holder. We use them for identification purposes only. This is not an official publication.

Motorbooks International titles are also available at discounts in bulk quantity for industrial or sales-promotional use. For details write to the Special Sales Manager at Motorbooks International Wholesalers & Distributors, Galtier Plaza, Suite 200, 380 Jackson Street, St. Paul, MN 55101-3885 USA.

On the front cover: The International 400 (later known as the W-400) was a strong workhorse for small-grain farmers in the mid-1950s and is a sought-after collector's item today. Andy Kraushaar

On the endpapers: The 600 gas and 600 diesel were not manufactured in great numbers and are sought after today. The diesels are more common than the gassers but are more complicated to restore. A collection with both is surely a sign of devotion to these powerful, fun-to-drive tractors. Andy Kraushaar

On the frontispiece: The Farmall 450 LP continued the line of LPG big Farmalls for IH. Always manufactured in smaller numbers than diesel or gas tractors, they're a great source of interest and conversation today. Andy Kraushaar

On the title page: The Farmall 450 LP had power 'til the cows came home. Andy Kraushaar

On the back cover: The Farmall 300 features live hydraulics and live PTOs, making it a superb choice for a collector or hobby farmer who is looking for a useful, interesting old tractor. Andy Kraushaar

Library of Congress Cataloging-in-Publication Data
Fay, Guy.
    Original Farmall hundred series tractors, 1954–1958 / by Guy Fay and Andy Kraushaar.
        p. cm.
        ISBN 0-7603-0856-X (hc. : alk. paper)
    1. IHC tractors--History. I. Kraushaar, Andy, 1953 - II. Title.

TL233.6.I38F39 2003
629.28'752--dc21

Edited by Lee Klancher
Designed by Chris Fayers

Printed in China

Contents

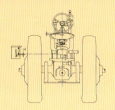

	Acknowledgments	6
	Introduction: How To Use This Book	7
Chapter One	General Information	8
Chapter Two	Cub and Cub Lo-Boy 1954–1958	14
Chapter Three	100, 200, 130, and 230	28
Chapter Four	Farmall 300, Farmall 350, International 300, and International 350	46
Chapter Five	International 330	72
Chapter Six	Farmall 400, International 400, International W-400, Farmall 450, and International W-450	77
Chapter Seven	International 600 and International 650	114
Appendix A	Serial Numbers and Production	130
Appendix B	Paint and Decals	138
Appendix C	Decal Drawings	139
	Index	160

Acknowledgments

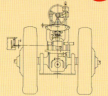

These books produce a lot of memories. You meet people, get to see collections, travel long hours on the road, and get to put up with Mother Nature. We did a lot of scouting for tractors at the Red Power Round-Up in 2000 at Sandwich, Illinois. The memorable part of that show was the great tractor turn-out in a scenic part of the country – and the rain. Got to meet Max Armstrong and be on the radio, which was pretty cool, but a lot of the show was spent slogging through the rain- Woody Woodyard from the ATIS Farmall List offered us a golf cart for the last part of the show, but me and Andy decided we'd get less wet if we just hoofed it, so went up and down rows I the pouring rain. Beats a drought, though.

Meeting Don Corrie at his place was a major thrill- I'm a big fan of the Wheatland type tractors, and Don's collection is pretty close to as good as it gets.

Darius Harms is someone every IH collector should meet. Great guy, didn't mind much when Andy locked him out of his truck. We asked him what he had in his collection. He asked us what we needed. Great collection.

John Sohner also had some great tractors, and a great place to shoot them. It was just one of those days where everything went right, and we got great photos.

We froze at Rick Wisnefshe's. Andy had to stop off and get a sweatshirt at a gas station, if I remember right. Thanks goodness Wisconsin tourist land sells a lot of sweatshirts!

We shot Kelly Gaston's tractor at the Sandwich show. Kelly at the time was just barely a teenager, but it was HER tractor. Of course Jack was great, too.

The Ripp boys had some great tractors, and a great background to shoot them against. And for once, it was a nice short drive north of Madison. I missed the Ramsey shoot- great tractors, too.

John Harper, Rich Saraga, and Steve Hile, CNH Global, Maureen Vlsac, Xerox. Case-IH still maintains a lot of the IH technical data. They sell tons of IH parts, and Rich and Mo are old Harvester employees. They still like me even when I screw up the microfilm machine again, and Mo lets me eat the in-house goodies….

Randy Leffingwell was good enough to help me out with some photos that we didn't have, and in general offer the support of a friend, even when the book drives me nuts . . .

The folks photographed:
Harms Farms (Darius and Family),
  St. Joeseph, Illinois
Jack and Kelly Gaston, Athens, Ohio
John Sohner Rockford, Illinois
Rick Wisnefske, Larson, Wisconsin
Mike Ramsey, Melrose, Wisconsin
Jim Ramsey, Melrose, Wisconsin
Roger Ripp, Waunekee, Wisconsin
Norm Ripp, Dane Wisconsin
And from Randy's photography:
Jay Graber, Parker, South Dakota
Bob & Mary Pollack, Dennison, Iowa
Don Corrie, Chenoa, IL
Tim Wehrman, Reese Michigan

Of course, thanks to the crew at Motorbooks- Lee Klancher, Darwin Holmstrom, and assorted crew, too. At the Wisconsin Historical Society, Lee Grady, Andy Kraushaar, Sherry Dolfin, Lisa Hinzman.

Guy Fay
February 15, 2003

Introduction: How to use this book

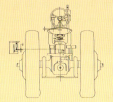

This book covers tractors built from the Fall of 1954 to the Summer of 1958, with the exception of the Super W-9 family tractors. This book covers U.S. production wheel tractors, including ag tractors and the industrial tractors based off those wheel tractors.

This book is arranged by tractor family—the 300 and 350s were very similar and had a large part commonality, the 100, 130, 200, and 230 shared a lot of parts, etc. A decal listing is at the back of each tractor, but due to the needs of printing this book economically, the decal placement charts (which are original IH drawings) are located at the rear.

A data chapter gives a variety of serial number, production number, and identification data for the tractors covered by this book.

Chapter One of this book covers some general history, and general information about common parts and restoration issues.

To use this book, the author would start with Chapter One, then go to the chapter covering the specific tractor you're resesearching. Use the data chapter and the decal charts as necessary. If you have a 330, read the 300/350 chapter as well. If you have a 300/350 Series, read the 400/450 chapter as well.

Dates

Dates given in this book for changes usually refer to the date the decision to make the change was authorized. Actual effective date varies due to manufacturing lead times, parts in stock, etc. When an actual date of serial number effective is available, it will hopefully be in this book!

Information Sources

Most of this information was taken from original IH specification lists (refered to by other manufacturers as bills of materials), product change notices, and engineering drawings held by CNH Global. It should be noted that errors in this original documtation, while very rare, still do happen. In general, the original production information should be more accurate than what is foud in a parts book, especially later edition parts books that often list later replacement part numbers, rather than the original production information. It should be noted than the information in this book covers only major changes, and then usually only if the involve exterior parts, due to the limited size of this book, as well as the time involved. There's still plenty of room for further research.

The Cub Lo-Boy was useful in buildings, around buildings, and anywhere else a small tractor with a low center of gravity was needed.

Chapter One
General Information

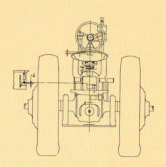

The International 100 next to a Cub. The Cub was roughly a ⅔-scale model of the Farmall Super A. The common genealogy is seen clearly here.

The peak of International Harvester (IH) tractor production was in 1951. Never again would IH make so many tractors. John McCaffrey, who replaced the ousted Fowler McCormick as the head of the corporation in 1951, was an old salesman who refused to believe that the glory days had passed. Although farm numbers were shrinking, acreage per farm was getting bigger, and so were the tractors. McCaffrey thought he could increase sales numbers by a system of "planned obsolescence," whereby farmers went out every year and bought the latest tractor. When some economic downturns, with accompanying low farm prices, hit the tractor business, things went bad for IH.

What was bad business back then is interesting collecting (and using) now. The IH tractors of the 1950s have a lot of variation to them. The Supers, which were mostly introduced in 1951–1953 and included the Super W-6 and Super M size, were followed by the TA tractors in 1954. Even the Supers saw interesting variations with the introduction of live hydraulics in the series and the Fast Hitch in the Super C.

In fall 1954, a whole new line of tractors appeared. Although this line shared ancestry with

The Farmall 400 was the row-crop tractor of choice for mid-1950s farmers, offering style, live hydraulics and PTO, and the Torque Amplifier, which made heavy tillage easier and offered a greater range of speeds.

the Super Series tractors and were made with the same basic tooling and many of the same components, the "Hundred" Series tractors had new sheet metal, different hydraulic systems in most cases, Independent PTO and Torque Amplifiers in the 300 size (replacing the Super H), and some other more minor differences. The Hundred Series was another evolution of the Super Series tractors, with the exception of the 300 Utility tractor, which was an IH answer to the Ford N Series tractors.

In 1957, more modifications of the Hundred Series tractor line occurred. The main differences from the previous tractors were the two-tone paint scheme and some hydraulic changes. The "50s and 30s" were again, a stretch of the same basic 1939 designs that had proven so popular for IH.

Notes and Philosophy of Restoration
The beauty of the antique tractor hobby is that there is no judging at shows. This tends to reduce the problems of people who, with large egos and checkbooks, just have to win. In the car world, these people bring cars to "prestige" shows in enclosed car carriers, then lay carpet from the ramp to the show area so that the tires (with the stickers still on them) never hit the ground, thus avoiding having to clean the tires. The cars are never driven, obviously. For tractors designed to pull manure spreaders, this seems a little excessive.

That being said, you don't have to restore your tractor like it says in this book. There are no rules saying that your tractor has to be restored accurately, or even inaccurately. A growing number of people are preserving tractors in their "working clothes" straight off the farm. Tractors in good, original condition document what they "really" looked like on the farm.

Sometimes, people become obsessed with restoring tractors to "factory" condition. However, with farm tractors other possibilities exist. Dealers installed a lot of equipment. Some tractors received LPG conversions before ever leaving the dealership. That tractor ran 40 years for a farmer in that condition. LPG conversions were an important part of farm history, and some deserve to be preserved as well.

Sometimes farmers or outside manufacturers added other equipment. The arc welder surely must rank as one of the top ten inventions that changed agriculture. If a farmer added some brackets, or made a homemade high-crop conversion, well, then at least consider preserving it. If a cane farmer took a cane tractor and put a gigantic "claw" cane loader on it, well certainly that deserves consideration as well. The point is, "factory" isn't the only way.

"As Maintained" Restorations

Many farmers maintained their tractors pretty well. Of course generators and starters got replaced, but farmers sometimes repainted and re-decaled their tractors as well. Sometimes a dealership did the job for free, kicking it in as part of an engine overhaul. That tractor you remembered driving as a child might have changed before you were born.

The decal service packages IH sold were not the same decals (either in appearance or number as what were put on a tractor originally.) The decals were whatever was current when the decal package was produced, not what was current when the tractor was produced.

Replaced components were often different from what was originally on the tractor. Starters and generators could be ordered from the dealership, bought from the nearest auto parts store, or ordered from a catalog like Central Tractor or Surplus Tractor Parts. These parts often came painted black. When IH bought these parts many were black but were then painted with the rest of the tractor on the assembly line. When Joe Farmer put on the new starter in the middle of planting corn, the last thought on his mind was painting it red. It worked just fine whatever color it was painted, and it was left that way.

Sometimes parts were sent in to specialty shops for rebuilding. In many cases, the parts came back painted black as well. This still happens today, so if you've got a part to repair that is original color, make sure to specify that it comes back that color as well.

Safety

One restoration standard that cannot be compromised is safety. While outside the scope of this book, safety is critical if you are going to work with, or show, the tractor. Accidents with antique tractors happen.

Mechanical reliability is imperative if you are going to have the tractor out in public. Here are some critical areas:

1. Brakes: Brakes must be fully operational, with sufficient amount of pad left. Brake locks must work and must not be blocked by sheet metal or other parts. Be careful when parking on an incline.

2. Fuel tank/cap: The fuel tank and fuel system shouldn't leak. The ventilation holes on the cap must be open and must not be blocked by paint or rust. Restorers of IH tractors should use the new "safety" gas caps available through Navistar for no charge. It may not be exactly original, but a burned tractor isn't original, either.

3. Exhaust manifold: An exhaust manifold leak aimed at the fuel tank can heat up the tank, causing venting or fire; not an obvious thing, but fires have happened.

4. Steering: Make sure the steering wheel is solid, especially if you're driving the tractor in road gear to a show. Make sure that there is no excessive play in the steering, as well.

5. Belt pulley: Make sure bolts and other hardware attaching the belt pulley are in good shape, and that if you are using the belt pulley, there are no cracks or chips. A runaway belt pulley can go for miles before it stops. Make sure the belt pulley won't damage a belt if in use; broken belts can penetrate skin.

6. Make sure that all hardware that is supposed to have safety wire/lock nuts has them, and all parts that are supposed to be properly torqued, are.

7. Make sure piston rings, valves, and mufflers are in good shape. I've been bathed in junk and oil coming out of mufflers before, and getting this stuff in an eye might result in a trip to the emergency room.

8. Make sure that tires are in good shape. A blown tire can take someone's arm off, if not kill him or her. Split rims are notorious. Make sure that lugs are fastened well on steel wheels.

9. Old wiring can cause fires. Botched rewiring can do the same thing.

10. Seats: Make sure that the seat pan isn't corroded, the attaching hardware is in good shape, and that the springs aren't cracked. Seats can, and do, fall off, usually when there's an operator in them. The tractor doesn't stop and neither does the disk behind you.

11. Hydraulic System: Make sure that hoses and other hydraulic system parts are in good condition. The pressures are more than enough to break skin. Leaking hydraulic hoses running alongside an engine—common for IH tractors of this period—might pose a fire risk.

12. Lights: If you are going to operate at night, the tractor's lighting system will need to be one hundred percent operational.

13. ROPS: A lot of people buy old tractors for mowing. They proceed to drive onto hillsides, roll over, and get squashed. Other people like to go to "antique tractor pulls" and soup up their tractors; the tractor does a back flip with the same result. A ROPS (Roll Over Protection System) is now

available for many older tractors; however, because of the nature of their construction, most of the tractors in this book *cannot* have a ROPS fitted. Don't think that you can just "weld up" a roll bar; tractor roll bars are finicky as far as materials and engineering. Doing it wrong can kill you when the transmission case shatters during a roll, dousing you with hot oil, or crushing you between the fender and bars. Contact your county extension agent for more information to see whether your tractor can use a ROPS.

14. Make sure your tractor is weighted properly, especially if you are plowing. If you have a tractor with the low-low first gear, IH specifications required front-end weights.

15. Make sure your clutch is fully functional. A non-disengaging clutch can be very dangerous, even through the tractor is still operational. People have been killed on tractors with clutches that drag enough to pin them against a low overhang, strangling them against the steering wheel, even though the clutch pedal was pushed to the stops.

Don't think that this list covers everything. Use your head, as well as everything else, to make an honest evaluation of your tractor. Talk to your local dealership, and talk to your county extension agent.

Other Information Sources

For people thinking about restoring a tractor, several different information sources should be found, begged, borrowed, or purchased. These include:

1. Parts books. Parts books show many of the parts as well as exploded views, which give a good idea on how to put a tractor back together. However, available parts manuals are often the latest revision only, and often do not show all superseded parts. A parts manual typically does not show all parts used throughout production. The only changes shown are usually where it would affect the purchase of spare parts. IH in the 1950s was intent on selling every last tractor they could, so some unusual parts or configurations can be found on tractors, but not in the common parts books.

2. Service Books. Engines and components had IH service manuals. Check Binder Books or other book suppliers to see what they have reprinted. In addition, I & T (Implement & Tractor) service manuals give actual disassembly and assembly procedures.

3. Operators Manuals. These manuals give operating procedures, as well as identifications and other information. Regular maintenance information should be of special concern to today's restorer. IH's operator's manuals were pretty good, and Binder Books, an authorized product licensee of Case IH, puts out very good reproductions.

The Farmall 450 gas provided power not only for row-crop operations, but also heavy tillage, haying, and other operations. That's why they call them Farmalls.

Parts Availability

Although some of these tractors are almost 50 years old at this writing, parts availability is good. However, you have to know if you are looking for a part to keep the tractor running, or to restore it to original condition. Your friendly local Case IH dealer will be able to supply a large number of the "keep it running parts" for these tractors. *Red Power Magazine* and the IH Collector's Club (IHCC) newsletter, *Harvester Highlights*, will have a large number of ads for specialty parts vendors. Don't forget to go to the IHCC's state shows and the national show, the Red Power Roundup, which is held annually, rotating around the United States. There's a Red Power Roundup held in Europe periodically, as well.

Demonstrator Programs

While most people are aware of the large midcentury sales promotions that had white-painted Cubs, Super As, and Cs produced by the thousands in 1950, most people are unaware of the less spectacular demonstration programs IH put on in the 1950s. These programs did not involve massive publicity or repainting the whole tractor, but did create some interesting items and possibilities.

The Broadview District Program

The Broadview District (comprising central Illinois) ran its own small demonstrator program in

Because the 650 had the white painted grill, it didn't receive the center piece of trim. These tractors have original-type, if not original, mufflers.

1956, using Farmall 450s fitted with power steering, draft control, and the two-point hitch. The tractors had a decal on the side naming the three features. The wheels on these tractors were painted yellow. About half the dealers in the district put on demonstrations with these tractors, which traveled from dealer to dealer during the program. There were no special serial number ranges for these units, and the painting of the wheels was probably done at the district, not the factory, level.

Rent-A-Cub

Extensive promotions were put on with the "Rent-A-Cub" program. The tractors were not (to the author's knowledge) prepared in any certain way as far as paint or equipment, although more fully equipped tractors were probably used in the program. Cardboard hood displays were used when the tractors were in the dealerships, and other special hang tags and display pieces were probably used as well. There were no special serial number ranges for these units.

Brass Tacks

Brass Tacks was a program that demonstrated well-equipped tractors. Feature stickers were located all over the tractor to identify these special features, such as hydraulics, TA, and power steering. The dealerships bought the tractors, demonstrated them, and got an allowance from IH to cover the depreciation.

Common Parts

Oil Gauges
Oil gauges on all Farmall tractors were changed in 1956. The older oil gauges had faces that showed the pressure range in shaded and white areas. The new gauges had faces marked in pounds of pressure. Cub, 100, and 200 tractors had faces reading from 0 to 40 pounds; 300 and 400 spark ignition tractors had gauges reading from 0 to 75 pounds. Diesels had a 0 to 45-pound range.

Heat Housers

Heat housers were an aftermarket accessory that dealerships sold. They were available from outside suppliers with their own brands, as well as IH. IH began offering its own line of heat housers around 1957.

IH referred to the housings as "Tractor Heaters." The attachment was a canvas duck housing that extended over the sides of the engine, and transferred the hot air from the engine and radiator back to the operator. On the IH versions, a tinted "triple thick" windshield and side housings kept wind from blowing directly onto the operator. There was no top or rear windshield, though the duck housing did wrap around the operator's position about equal to the top of a deluxe seat. The main body was in Harvester blue, with red trim and the IH symbol and "Farmall" stenciled on the side. The "heat flaps" that extended over the engine could be rolled back on warm days. The frame of the housing was made of angle iron, and the seams in the canvas were all double stitched.

Fuel Caps

Fuel caps have been one of the most controversial parts of the IH tractors of this era. Currently,

International Truck and Engine Corporation offers a free replacement gas cap for all gasoline tractors of this era, (check dates). The new caps offer improved venting to handle today's more modern, volatile formulations of gasoline. Anyone wishing to receive a new cap should to write to:

International Truck and Engine Corporation
Box 10088
Fort Wayne, IN 46850

Please remember to send along the serial numbers of the tractors for which you need caps. They may have a form to fill out (which is available from many sources, including Binder Books). The caps may take some time to arrive. However, if you are using your tractor to any degree, the cap may mean the difference between a nice restored tractor and a real problem.

The standard fuel caps were the flat caps with ridged edges. However, a popular option was made available in the mid-1950s. The new fuel caps had a gas gauge built into the cap, with a cork float assembly supported by two pieces of rod. The cork moved up and down the rods, and rotated a twisted piece of metal. The twisted piece then rotated the needle in the gauge to reveal the fuel level. Different fuel gauges were used by fuel tank capacity, with some gauges being available for older tractors as well. The gauges have the IH logo on the bottom of the gauge under the needle pivot, and a scale that wrapped nearly around the gauge, registering in quarters.

The part number of the gauge is located under the IH logo:

Part 365482R92 was used on the 100, 130, 200, and 230 family tractors, as well as the A, B, and C families dating back to 1939.

Part 365495R92 was used on all the 300 and 350 Series tractors, and on the 400 and 450 diesel tractors. Farmall 400 and 400 HC non-diesels under serial number 38333 and W-400s with serial numbers below 3596 also used this cap.

Part 365497R92 was used with the big fuel tank 400 tractors (F-400 and F-400 HC serial 38333 and up, and W-400 3596 and up) and the 450 serial non-diesel tractors.

Three-Point Hitch Adapters

Three-point hitch adapters were available from IH. The adapters were available for 230, 350, 450, and International 300 and 350 Utility tractors with Fast Hitch. IH advertised that three-point hitch implements would work as well or even better than normal because of features inherent in the Traction Control and Fast Hitch designs

Tires

Identifying what brands were supplied for which models proved impossible at the time of this writing. A chart was discovered that included the names of the tire suppliers IH used, including the trade names of the respective tires (See Table 1.1).

Table 1.1
Tire Codes and Names

Code Marking	Tire & Rim Association and Rubber Manufacturing Association Name	Goodyear Trade Name	Firestone Trade Name	B.F. Goodrich Trade Name	U.S. Rubber Trade Name
Front Tractor Tires					
F-1	Single Rib	Single Rib	Single Rib Guide Grip	Skid-Ring	Mono-Rib
F-2	Triple Rib	Triple Rib	Guide Grip	Multi-Ring	Tri-Rib
F-3	Industrial Rib	Multi-Rib	Rib Tractor	Ribbed	Multi-Rib
Rear Tractor Tires					
R-1	Farm	Super Sure Grip	Champion Ground Grip	Power Grip	Grip Master
R-2	Cane and Rice	Special Sure Grip (R-2-0) Rice Special (R-2-C)	Champion Spade Grip	Special Service Canefield Special Service Ricefield	Cane and Rice
R-3	Industrial	All Traction, All Weather, Industrial Sure Grip	All-Non-Skid	Tractor Grader Lug Type All Purpose, Sand Grip	Industrial Grader,
G	Garden Tractor	Sure Grip, Super Sure Grip	Garden Tractor	Traction Implement (Conventional), Super Hi-Cleat Garden Tractor (Wide Base)	Ground Drive

*Tractors shipped overseas, especially to rubber-producing areas, were often shipped without tires or rims, then were fitted with local brands

Chapter Two
Cub and Cub Lo-Boy 1954–1958

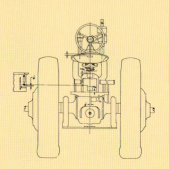

In late 1954, the Farmall Cub was restyled to match the rest of the IH line. A one-point Fast Hitch became available at the time, but this tractor doesn't have one because it has the rare "High and Wide" attachment.

The Farmall Cub received several changes in late 1954. The most visible change was the restyling, which involved a new hood and fuel tank assembly, new radiator grill and screen, and IH and Farmall Cub emblems instead of the old decals. The changes required the design of a new under-axle muffler and exhaust attachment. The air cleaner was modified to use a shorter intake air stack, and a new fuel tube was added to clear the new exhaust. These changes took place with serial number 185001, the start of the "Hundred Series" styling. This styling (including the addition of the white grille and side decal in 1956) lasted until the introduction of the 1958 style—matching the 40 and 60 Series tractors. The last serial number was 210000.

The 1955 introduction of the International Cub Lo-Boy filled a long-term need in the IH industrial line. The discontinuance of the I-12 in 1940 left industrial dealers without a small tractor capable of working in the very tight clearances of factories and other facilities. While Hebard Shop Mules could perform some of those duties, it went out of production in the early 1950s. The Industrial Power Division had experimented with an Industrial C in the late 1940s, but nothing came of the project. Attention then turned to the Cub. Putting the Cub through roughly the same

How the tractor got the extra height on the rear is obvious here—larger rims and tire equipment and extenders to match them up with the wheels. Somebody's obviously repainted the front wheels.

program as the C and rotating the final-drive drop housings, a prototype Cub Lo-Boy was built in 1951 or 1953.

A non-adjustable front axle was used, with tread adjustment made at the front and rear wheels. The resulting tractor was 8 feet long, had a turning radius of 8 feet, and a height over the steering wheel of 55 inches. Total ground clearance was 10¾ inches under the transmission case. Operating weight was listed at 1,640 lbs. The small-size tractor soon found favor in a wide variety of operations where small sizes or a low center of gravity were critical. Government highway departments bought many of the tractors for roadside mowing; some were even purchased with rear dual tires—not for the increased traction, but because the extra wheels allowed the tractor to work on steeperslopes.

In the early years of International Cub Lo-Boy production, the Lo-Boy name was only used in the United States and Canada. Outside the United States the name was "McCormick International Cub Special in certain markets."

The two smallest tractors in the Farmall line didn't change names or major mechanical features for the 1957 sales season. Major changes included

Table 2.1
Cub Kinds and Codes

801 Code 1	801 Code 2	500 Code 2	500 Code 3	500 Code 4	501 Code 1	501 Code 2	501 Code 3	Part Number	Description
x	x	x	x	x	x	x	x	356098R91	Air wheel & 2.50Cx12DC Rim
x	x	x	x	x	x	x	x	355052R91	4.00x12 4 ply F-2 Tread tire and tube
x	x		x	x		x	x	356102R91	Rear wheel & 7-24 Rim
		x			x			356100R91	Rear Wheel & 5-24 Rim
		x			x			358594R91	7-24 4 Ply R-1 Tread Rear Tire & Tube
x	x		x	x		x	x	356110R91	8-24 4 Ply R-1 Tread
			x	x		x	x	352514R97	Electric Starting & Lighting Attachment (with Battery Ignition)
			x	x		x	x	357543R91	Deluxe Type Upholstered Seat Attachment
				x			x	362877R91	Fast hitch Attachment
				x			x	360711R92	Touch Control Attachment
				x			x	351389R91	Adjustable Front Axle Attachment

*The export tractors usually had the same equipment as the domestic kinds and codes, but also included oil filter element packages.

The hub extensions for the rear wheels are shown here. This tractor has a plain drawbar on it—the Fast Hitch couldn't be used because of the tractor's extra height.

new white paint on the radiator grille, a new white hood sheet nameplate background decal, and a new IH hood emblem with a Harvester Red background. The tractors were restyled for the 1958 season to match the new models coming in across the IH ag tractor line, including a new barred grille. The last serial on the Farmall Cub before the 1958 changes was 210000, while the last Cub Lo-Boy was 10000 (See Table 2.1).

Kinds and Codes
Kind 801 Cub Domestic Sales Code 1 (post-1956 version)
Kind 801 Cub Lo-Boy Domestic Sales Code 2 (post-1956 version)
Kind 500 Farmall Cub (1955–1956 version) Codes 2, 3, 4 Domestic (apparently no Code 1)
Kind 501 Cub Lo-Boy (1955–1956 Version) Codes 1, 2, 3

Engine
The engines were the familiar C-60 used in the Cub since 1947. An under-axle muffler was standard on the Lo-Boy. A vertical muffler was available as a special attachment. Engine speed on the Cub was originally 1,600 rpm, but the speed was increased in early 1956 to 1,800 rpm. The change increased power to 9.4 on the drawbar and 10.5 on the belt. Cast iron flattop pistons were used in this era of Cubs and Lo-Boys. Later, Cubs used aluminum-dome top pistons that could be found in the older tractors as replacements. Valve rotator and high-altitude cylinder heads were available as attachments on both Cubs and Cub Lo-Boys

"Super Cub"
Readers may be aware that IH made Super Cubs at their factory at St. Dizeir, France, that were never emported to the United States. However, it seems that IH may have unofficially built a few higher-horsepower Cubs that were referred to as "Super Cubs" at Louisville, Kentucky. A single reference (See Table 2.2) does refer to an engine combination called the "Super Cub," with the Super Cub in quotation marks. The engines used a cylinder head usually found with a unique manifold in U-1 power-unit engines, 55 hay baler engines, and 64 Harvester Threshers.

Cooling
Radiators had two sources: IH and Moline.

Table 2.2
Cub Piston/Cylinder Head/Compression Ratio Chart
(Beginning in 1952, Extending to About 1956)

	Piston	Cylinder Head Manifold	Compression Ratio
Cub	251240R1 251232R1	251228R1	6.5:1
Cub 5,000-foot altitude	251240R1 251232R1	351779R1	7.45:1
"Super Cub"	251240R1 757338R1	355691R1	7.0:1
"Super Cub" 5,000-foot altitude	251240R1 757338R1	351779R1	7.45:1

Starting Crank
Starting cranks were still available for tractors built without starting systems. The crank was inserted into the fan drive pulley, which was the same for all models.

Exhaust
Several options were available for Cub and Lo-Boy exhausts. The standard exhaust for the Farmall Cub and the International Cub Lo-Boy was the under-slung exhaust. The muffler as supplied from the factory had an IH logo and part number stamped into the muffler shell. The under-slung muffler was 363116R91, and was made by either Donaldson or Mac Kenzie. Mufflers were made of aluminized steel, were 3 inches in diameter, and had the IH part number and other information stamped into the shell. Hood and fuel tank assembly 362369R91 (later R92 and R93) was used, which did not have a muffler hole in it.

Available as a special option for both the Farmall Cub and International Cub Lo-Boy was a vertical muffler. These tractors used a hood and fuel-tank assembly that had a hole in the hood for the muffler. The muffler was part 351436R92 (356003R91 was the same muffler from a different source). Donaldson, Mac Kenzie, and Hayes made the mufflers.

Spark Arrester
Curiously, IH did not offer spark arresters for Cubs built after Serial 185,000, including the 1954–1958 Cubs. IH did offer spark arresters for earlier tractors for field application through the 1950s.

Air Cleaner
Donaldson and United Specialists air cleaners were both used in production on Farmall Cubs and International Cub Lo-Boys. The Donaldson air cleaner, 350747R92, had a more rounded top, while the United Specialties air cleaner, 350749R92, had a flat top. Both had decals in the same place and were supplied for assembly line use in a red oxide primer; spare parts stock were provided to IH with red paint. Air cleaner decals were available in English, Spanish, and French on each, supplied by the air cleaner manufacturer.

Electrical
A magneto ignition was standard on the Cub and Cub Lo-Boy. Starter and lights were special attachments, as was the battery (distributor) ignition. Most tractors of this era were shipped with battery ignition, due to dealers buying tractors as "sales packages" that had a more modern ignition. Starting and lighting systems were available as special attachments for both battery ignition and magneto ignition tractors, with factory or field application. Starting in January 1957, the systems were also available as parts accessories.

The electrical systems on these tractors used a Delco-Remy generator. The voltage regulator was separate from the generator.

Later tractors may have the generator replaced with a more modern unit that features a voltage regulator mounted directly on the generator. The replacements were known as the "universal generator package." Generator pulleys changed at serial number 196757, from 358212R1 to 364907R1. The later pulley had a center that did not protrude on the radiator side.

To get the extra height for the attachment, taller spindles were used. The text describes the entire package.

The Farmall Cub received the new two-tone color scheme to match the rest of the line in 1956. This tractor is the normal height and width. The grill is, of course, painted white. The seat is the correct color for this vintage—the silver is from the waterproofing compound used at that time.

enclosed in its own little bubble housing, and had a ruby lens to change the light to red. The smaller light was mounted on the main taillight body.

Some export tractors received the older-style non-sealed beam headlights, which also used a "double lead" wiring harness. The notation for the attachment in the specification list states that these tractors were for shipment to Sweden, which legally required the older-style lights.

Battery
Globe Union and Auto-Lite batteries were used.

Ignition Switches
Cubs and Lo-Boys of this era had the push/pull-type switch, although later Cubs in the 1960s had turnkey switches.

Wiring Harness
The main wiring harness, 363507R91, carried most of the wiring. It used a woven covering. A separate, rear-lamp cable, 363508R91, was used. A battery-to-ground cable was used—69510 DA or 365048R91 (both rubber covered)—until April 27, 1955, when the latter was eliminated. The generator A terminal-to-regulator wire, 358112R92-365050R91 (FTC12472), and the generator F terminal, 358113R92-365051R91, also were eliminated April 27, 1955. The parts eliminated were made of materials that were not in critical demand during the Korean War period, and were eliminated when material shortages were over. Headlight grommet 365844R1 or 120302 was used where the headlight cables passed through the hood. Grommet 52118-D (rubber) was used where the cable passed through the battery box.

Starter
Starter motors used with this era Cub had a die-cast starter switch mounted directly on the starter. A Delco-Remy 6-volt starter was used.

Lights
Headlights and rear lights were identical on these years of Cubs and Cub Lo-Boys. Standard was sealed-beam 6-volt units, part number 357884R94 made by Guide Lamp. The rear light bracket on a Cub Lo-Boy had an extension to raise the level of the light (some Cubs might also be found with this feature).

A combination rear light and taillight was available as special equipment. The taillight was

Electric Breakaway Connector and Safety Lamp
An electrical, breakaway connection socket was available as a special attachment for the Cub and Cub Lo-Boy. The connectors were used with a safety lamp (also a special attachment) that could be put onto a trailing implement or wagon for use at night. Guide Lamp made the lamp under its number 897104 (IH Package 363914R91). The breakaway connector was located on FZG-1580. The Electrical System Caution decal (6 volts), 1001725R1, was used with this system.

Spark Plug Wires
A spark plug wire bracket was located on the top of the engine. A rubber grommet in the bracket

Part of the two-tone color scheme was the white decal that goes behind the nameplates on the side of the hood. This Cub has the underslung exhaust. Note that the front wheels and rims are all red on the Cub.

Photographs of tractors newly produced at Louisville show that wheel nuts and bolts were left unpainted (but were cadmium plated). The Cub used the small C-60 engine that, unlike the larger IH tractor engines, didn't use sleeves. The serial number plate is just above the front axle.

kept the wires from rubbing through the insulation. A cable tie was used. Nipples with the IH logo molded into them were used on the magneto end of the spark plug and coil wires.

Magneto
An IH J-4 magneto was used.

Chassis
Serial Number Plate
The serial number plates on all tractors were modified by a decision authorized March 7, 1955, which eliminated the maximum speed idle number and the overload warning.

Steering
The Cub and Cub Lo-Boy used essentially the same steering apparatus. All Cubs and Cub Lo-Boys below serial 4802 used a straight steering gear arm. Lo-Boys 4802 and above used a steering gear arm that curved down, while Cub Lo-Boys fitted with a 105 mower used an arm with two 90-degree bends.

The standard steering wheel was a 15-inch-diameter type made by either Sheller (60069-D) or French & Hecht (46512-DA). Cub Lo-Boys with a 105 mower used a heavier built, 18-inch steering wheel with an insert in the middle with "IH" on it.

Clutch

Both Auburn and Rockford clutches were used. The Auburn was officially standard and did not receive a letter code suffix, while the Rockford featured a J. Most clutches in the tractors of this era seem to be a Rockford and are generally considered to be a bit heavier duty than auburn.

Transmission

The Cub used a three-speed transmission. When the change was made in engine rpm from 1,600 to 1,800 in early 1956 (serial 192113?), the transmission gear ratio was also changed. The new speeds were:

1st gear 2.435 miles per hour
2nd gear 3.246 miles per hour
3rd gear 7.304 miles per hour
Reverse 2.713 miles per hour

Tractors without a PTO were fitted with a flat oval plate covering the rear of the transmission housing.

Table 2.3
Front Rim and Disc Attachments

Attachment Number	Rim Size
356098R91 (Cub and Lo-Boy)	2.50X12DC
356099R91 (Cub Only)	3.00x15DC

Table 2.4
Front Tire Attachments

Attachment #	Tire Size and Tread	Plies	Tread	Used with Rims
355052R91	4.00x12	4	F-2	2.50Cx12
356106R91	4.00x15	4	F-2	3.00Dx15
364784R91	4.00x12	4	I-1	2.50Cx12

Attachment 364784R91 was added August 12, 1955.

Table 2.5
Rear Wheel and Rim Attachments

Attachment number	Rear Wheel	Rim Size	Rim Manufacturer	Tire Size	Plies	Tread
356100R91	351083R1	W5-24	351084R1	7-24	4	R-1
356102R91	351083R1	W7-24	351428R91 Goodyear Firestone French & Hecht Electric Wheel	8-24 9-24	4 4	R-1, R3 R-1
356104R91	351083R1	W5-30	355076R91	6-30 7-30	2 4	R-1 R-1

35100R91 and 356102R91 used with both Farmall and Lo-Boy. 356104R91 Farmall Cub only. Possibly other tires were used with 356104R91. Screws and nuts were cadmium or zinc plated.

Table 2.6
Rear Tire Attachments

Attachment #	Tire Size	Plies	Tread	Used with Rims
358594R91	7x24	4	R-1	W5x24
356110R91	8x24	4	R-1	W7x24
356111R91	9x24	4	R-1	W7x24
356112R91	6x30	2	R-1	W5x30
356113R91	7x30	4	R-1	W5x30
356137R91	8x24	4	R-3	W7x24

The gear shifter actually varied by the seat fitted on the Cub. Farmall Cubs with the regular seat received a straight gear shaft lever. Number 351548R11 replaced 364067R1, which was only used from March to June 1955 on tractors with deluxe seat attachments; 351548R1 was the original part and IH actually switched back to it.

All International Cub Lo-Boys, and Farmall Cubs with the deluxe cushion attachment, received a curved gear shifter. The lever housings (364065R11) and swivels also differed from the Farmall Cub.

Rear Axle

The rear axle housings and right rear axles differed between the Cub and Cub Lo-Boy tractors. The left-hand axles were identical (and were used on the Lo-Boy, as were the right rear axle). Cubs fitted with the high-clearance attachment were fitted with rear axle extensions, which were a spindle with two flanges on each end to mate with the axle and wheel bolts.

Front Axles

Both the Farmall Cub and the International Cub Lo-Boy had a non-adjustable axle as standard, with an adjustable front axle available as an option factory or field. The spindles were the same on the adjustable and the standard axle on both tractors, but the Farmall spindle was much taller than the Lo-Boy spindle. All axles were tubular in section. The Cub adjustable front axle treads went from 40⅝ inches to 56⅝. The Lo-Boy adjustable front axle originally had treads of 43 to 55 inches, adjustable in increments of 4 inches, but this changed to 39 to 55 on August 17, 1956. The new axle extensions had five holes instead of four. The 39-inch minimum tread was for use only on level ground. On rough ground, the left extension had to be set at a minimum of 21 inches from the pivot pin to avoid interference between the tire and steering gear in a left turn.

Wheels and Tires

IH officially did not consider wheels and tires to be "regular" equipment—all were ordered separately, with different options, listed in Tables 2-3 to 2-6. However, when an order for a tractor did not specify which wheels and tires were to be fitted, IH supplied a standard set of equipment. For the front wheels and tires, a 2.50Cx12 DC rim with 4.00x12 4-ply F-2 tread tires were supplied for both Farmall Cub and International Cub Lo-Boy. For the rears, the Farmall Cub orig-

A common use for Cubs today is to power belly mowers for people with large lawns. This one has an aftermarket seat cushion.

This tractor is fitted with the one-point Fast Hitch. The Fast Hitch was IH's unique answer to the three-point hitch. The easy-to-use Fast Hitch required the operator to back into the one prong on the implement with the receiver to automatically latch. The Fast Hitch was done in by the 14-year headstart of the three-point hitch.

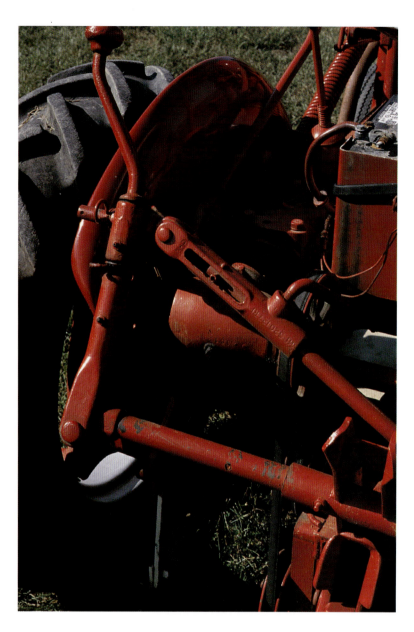

The Fast Hitch did have its idiosyncrasies. This is one of the adjustment points. The receiver is seen at the lower left.

mounting brackets and swinging drawbar were different for the models due to height. On the Farmall Cub, the drawbar attachment to the tractor was on the top side of the drawbar, with the other half of the tongue welded to the bottom of the bar; on the Lo-Boy, both pieces were welded to the underside of the bar.

Seat

The Cub used a post-type seat support; the Lo-Boy used a mounting bracket. The standard seat on the Lo-Boy and the Cub was 351880R92: a metal seat pan made by Milsco. Optional were a foam-rubber padded, pan-type seat; a deluxe cushioned seat with padded backrest; and armrest pads for the deluxe seat.

Two different, cushion-upholstered seats were used in production, made by Milsco and Bostrom. These seats were originally supplied in silver (coated with Koroseal), but subsequent parts supplies were available in off-white. For a while after May 1955, seats were sold with a tag saying "Genuine Foam Rubber" attached at the center seam. In late May 1956, the filling become polyurethane, rendering the tag obsolete. These seats were available factory or field. See the parts accessories section at the end of the chapter for the optional seat pad description.

The "deluxe" cushioned seat, optional on the Lo-Boy, was made optional for the regular Cub in early March 1955, with or without arm pads. Both Milsco and Bostrom made the special-attachment armrest pads, part 363740R91 (the attachment was 363739R91). The fabric was coated with silver Koroseal, while post-1958 service parts were imitation black leather. The seat itself had tube arms on both sides that curved around in back to support a back cushion; titled "rest assembly" in the parts books, part 363219R91, also had a white cotton duck cover coated with silver Koroseal (again, imitation black leather after 1958). A square steel frame supported the arms and the square, thick, seat bottom cushion. Both Milsco and Bostrom made the deep cushion on the bottom. As noted in the transmission section, a different, curved, gearshift lever and a housing were fitted on the Cub when the deluxe cushioned seat was fitted, to prevent seat interference with shifting.

Toolbox

Toolboxes were included in the deluxe seat assembly and were standard on all Lo-Boys. The Farmall Cub with regular seat had a slightly

inally was shipped with a W7-24 rim and an 8-24 4-ply R-1 tread tire. On the Cub Lo-Boy, a 5-24 rim and a 7-24 4-ply R-1 tread tire was used. On March 31, 1958, the 7-24 rim and 8-24 tire were made the standard on both tractors.

Attachment 356113R91 was eliminated September 26, 1955, but was reinstated November 10, 1955. Cancelled as part of an industry-wide effort to eliminate rarely used tire sizes, demand was later found to be enough to warrant keeping the tire in production.

Hitch

A standard drawbar was regular equipment. A swinging drawbar attachment was available for factory or field application. The U-bar drawbar was the same part for Cub and Lo-Boy, but the

The Cub Lo-Boy, originally introduced in 1955, was updated with the two-tone color scheme in 1956. Owner Kelsey Gaston is in the seat.

different toolbox, 351014R91, which had the bottom of the toolbox located not at the ends of the sides, but a couple inches up.

Sheet Metal

The hood and fuel-tank assembly was restyled from the previous styled Cubs. The same assembly was used on Cub and Cub Lo-Boys. The new style eliminated the horizontal ribs, and instead used a recessed area to fit new trade name emblems (and on the 1956–1958 tractors, the white side decal). The radiator grille was new, with horizontal bars instead of plain wire mesh. There was a galvanized 26-gauge steel wire 5 x 5 mesh screen inside the grille. The bar-type radiator grille and the fuel-tank and hood assembly was used for parts replacements for older Cubs, therefore, pre-1954 tractors may be found with the 1954–1958-style grille and hood. Separate hood and fuel-tank assemblies were used for export. The primary difference was that the nameplate holes were drilled differently for the foreign nameplates.

An IH hood emblem went on the top front of the hood where it curled over the radiator. The Farmall Cub had two emblems on the side: a small, "McCormick" emblem (in North America) and a large "Farmall Cub" emblem underneath. A different hood and fuel-tank assembly was used for tractors with a vertical exhaust (the vertical muffler hoods had muffler holes).

The International Cub Lo-Boy had one "International Cub" emblem and a "Lo-Boy" emblem running down the side of the radiator grille. Some International Cub Lo-Boys sold overseas were not allowed the Lo-Boy name, due to other firms using that name. Instead, these tractors had a "Special" decal in the same location. The screen sides without the holes for the nameplates were taken from Farmall Cub production. Additionally, some Farmall Cubs were sold overseas with "International" nameplates, while some International Cub Lo-Boys were sold overseas with the "McCormick" nameplate.

The Lo-Boy had a lower center of gravity. The final gear drives were rotated so they were not pointed straight down, but rather sideways, lowering the rear end of the tractor. The front spindles were shorter.

23

A lower tractor could handle steeper slopes (especially for roadside mowing), but could also go into areas with limited overhead clearance, including orchards. The grill was painted white on the 1956 to 1958 model, but the 1955 and early 1956 tractors had a red grill like the rest of the line at that time.

Rear Fenders and Support Plates
The fenders were the same between the Cub and Cub Lo-Boy. However, screws and bolts were different lengths—both between tractors, and between Fast Hitch and non-Fast Hitch versions. The Cub Lo-Boy also had two fender support plates, and a rectangular plate, right-hand fender brace.

Hydraulics
Touch Control
Both the Cub and Cub Lo-Boy had Touch Control available as a special attachment; it was not standard on either tractor. Of course, when dealers ordered for inventory, they almost always ordered tractors with Touch Control, with dealer "sales packages" set up that way. Touch Control packages were also available for in-the-field application for tractors originally shipped without hydraulics. When Touch Control was added—either factory or in-the-field—a hydraulic lift, oil-level decal was added on the fuel tank just above the filler. Decal 1001348R1 was used (English, 1001349R1 Spanish, 1001350R1 French).

Rockshaft
The rockshafts for Touch Control on the Hundred Series Cubs were fitted with helper spring arms; older Cubs did not have helper springs. IH offered a service kit to retrofit the newer-style arms and springs.

One-Point Fast Hitch
One-point Fast Hitches were available for both the Cub and the Cub Lo-Boy, beginning with the 1954 restyled models. The hitches and mounted equipment were different between models because of different heights of the tractors. The one-point Fast Hitch had a flip-over drawbar for pulling wagons or other implements.

Fast Hitch packages were also available for field installation. The tractor of course had to have the Touch Control hydraulic system installed. Fast Hitch packages were available for pre-1954 tractors as well. Installation involved reworking the fenders on the older tractors, as well as drilling some attaching holes on the platform. Knob 53717 DA (rubber) was used on the depth-adjusting screw on both the Cub and Cub Lo-Boy tractors.

High and Wide Attachment
A high-clearance, wide-tread attachment was available for the Farmall Cub, both from the factory and as a field attachment. There were two different factory attachments: 355061R91, which included 2-ply, 6-30 R-1 tread rear tires; and 355472R91, which included 4-ply, 7-30 R-1 rear tires. Both attachments used 4.0x15 F-2 tread front tires. Attachment 355062R91 was available for Farmall Cubs in the field that were already equipped with the adjustable, front-axle attachment 351389R91 (factory high and wide tractors came with the adjustable axles). The 3.00x15 –drop-center front rims were used, as were W5-30 rear rims on both factory attachments. Both front and rear axles had extensions to move the wheel/tire assembly out from the sides of the tractor, increasing stability.

Exhaust Valve Rotators

Exhaust valve rotators were available for Cubs and Lo-Boys of this era.

Power Takeoff

The increase in engine speeds in 1956 also resulted in an equal increase in PTO (power takeoff) speed to 1,800 rpm. The change in the PTO speed resulted in the shaft being grooved for a retaining snap ring that was also used for repair parts for earlier tractors.

Belt Pulley/PTO Unit

Cub and Cub Lo-Boys had combination PTO and belt-pulley assemblies available as special attachments, factory or field. International Cub Lo-Boy tractors that were ordered from the factory with Fast Hitch and the combination units did not have the belt pulley assembled on the PTO unit (it was shipped separately). A PTO-only assembly was also available as a special attachment, factory or field.

The PTO attachment was non-standard, revolving the opposite direction of the standard American tractor PTO. At least one company (Hub City) built attachments to reverse the direction of travel.

When the Cub received the increased rpm engine (from 1,600 rpm to 1,800) in 1956, the standard belt pulley diameter was reduced to 7⅝ inch, to hold the belt speed roughly equal to the previous speeds. Nine inch diameter by 4¾-inch face pulleys (351284R1, gray iron) were used by the slower speed Cub and Lo-Boys. The 7⅝ by 4 3/4 pulleys (351285R1, gray iron) were used by the higher speed Cub and Cub Lo-Boys as standard for tractors with the belt pulley/PTO special attachment. Optional belt pulleys were available for both the fast and slow versions—the slow version had a 7 5/8-diameter (351285R1) and a 6-inch diameter (354905R1) grey iron pulley available, while the high-rpm tractors had a 9 inch (351284R1) and a 6 inch (354905R1) available. All had 4¾-inch faces. Several different, gray iron weights were riveted onto the inside of the rim to balance the pulley.

Hour Meter

A Hobbs hour meter attachment was available for Farmall Cubs that featured electric starting and lighting attachments. The attachment was carried under 356735R91.

Front and Rear Wheel Weights

Front wheel weights (part 351371R1) that weighed 26 pounds each were available, and could be used up to two to a side (restricted to 12-inch wheels until May 1957). The weights were made of gray iron. In January or May 1958, 50-pound split-type weights for inside the front wheel that could be used with either one or two 26-pound outside weights were added. Rear wheel weights (351374R1) that weighed 150 pounds each were available, usable in one or two weights per side.

The manifolds on these tractors were originally painted red, but the paint quickly burned off in use. The Lo-Boy model nameplate runs vertically down the side of the grill.

Right: A view from the operator's position showing the tractor's excellent visibility. The steering wheel is properly detailed (red spokes, black rim). The center nut may have been left unpainted in production.

Far right: A view of the final drives of the Lo-Boy—they were rotated 90 degrees forward, lowering the tractor by several inches.

Fast Hitch Cushion Spring

A cushion spring for the Fast Hitch was available originally as a field attachment for the Cub, but was offered as a factory attachment starting August 10, 1955. The cushion spring later became a parts accessory. The thick coil spring was also available on Lo-Boys.

Horn

An electric horn assembly was available from IH for the International Cub Lo-Boy; probably made first available (factory or field application) in August 1955. Delco-Remy made the equipment. The switch was 364512R91, furnished in red enamel, and was similar to Delco-Remy part number 1996027. The horn switch was a push button with a rubber cap.

Toolbox

Two different toolboxes were used: one for the standard seat on the Farmall Cub, for the deluxe seat on the Farmall and for all International Cub Lo-Boys. The second toolbox had a cutout on the bottom.

Tractors shipped overseas had a lubricator compressor (otherwise known as a grease gun) as standard. A spark plug wrench (46190-DC, ? inch) socket with handle (28252-D and 25466-D) were the only other tools shipped with the 1955- and 1956-style Cubs. These tools were eliminated when the new spec list page was issued May 7, 1956.

Mounting Step

An optional mounting step was available from IH to assist operators in getting on the tractor. This part was carried under 368678R11 and was available on Cub and Cub Lo-Boys probably starting in the 1956–1957 time period.

Pneumatic Tire Pumps

Schrader and Engineair Spark Plug-type tire pumps were available as special attachments for both Cub and Cub Lo-Boy tractors.

Decals

1000867R1 lighting switch (tractors with lighting systems only)
1000636R5 PTO warning (English, 1000772R5 Spanish, 1000636R4 French) (when equipped)
1001718R1 "Made In the United States of America" (all tractors shipped out of the United States)
1001811R1 "Special" (Lo-Boys exported)
01001604R1 patent marking changed to R2 October 18, 1956, R3 July 23, 1958

A rear view of the Lo-Boy. This tractor has the rear rockshaft installed. Original seat cushions were silver, but later IH production (and spare parts) of the identical cushion was black.

1001130R5 oil filter instruction (English, 1001131R5 Spanish, 1001132R5 French)

1001182R1 warning, drawbar and front pull hook (English, 1001184R1 Spanish, 1001185R1 French)

1000987R2 cylinder head breather (English, 1000988R1 Spanish, 1000989R1 French)

1000704R4 warning, brake (English, 1000767R4 Spanish, 1000819R3 French)

3750011R1 left-hood sheet-side decal (added September 5, 1956)

3750012R1 right-hood-sheet side decal (added May 7, 1956)

2750194 Cub trademark (added May 9, 1957)

Parts Accessories

In May 1956, IH changed the way it handled certain accessories. At first called "Special Service Items," these parts were handled under the term "Parts Accessories" after January 1957. These parts could be ordered new on farm tractors. When farmers wanted to buy them to put on tractors already delivered, the parts were sold from parts stocks, and were not officially attachments, thereby reducing the inventory count. Some of these parts were sold directly from displays in the dealers and were among the most commonly added (or replaced) parts after the tractor was sold. These parts were purchased, not produced, by IH, and included the following on the Cub and Cub Lo-Boy:

Combination rear and tail lamp 361985R93
Detachable seat pad 351438R93
Detachable seat pad 359483 R91
Upholstered seat 357518R91/2/3
Cushion upholstered seat 364398R91 (Lo-Boy only)
Extension cable complete 365444R91
Radiator pressure cap 365892R91

Chapter Three
100, 200, 130, and 230

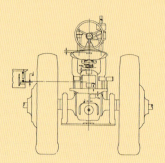

This is an example of a restored International 100. The International is pretty rare, although how rare is a bit of a mystery—it was at first a separate model, but later became an option to the Farmall. This one has the correct silver-color seat.

The Super A and Super C were updated to the Farmall 100 (including 100 high clearance and the International 100 industrial) and 200 with a mild restyling and new hydraulic systems in late 1954. The major changes on the 100 and 200 chassis were a new radiator grille, hood, and nameplates instead of decals to identify model and make. Fast Hitch, available on the Super C since 1953, was extended to the 100, which got the one-point Fast Hitch that used some linkages similar to the abortive Frameall of the mid-1940s. In the case of the 100 and 200, the Fast Hitch worked with the old Touch Control hydraulics.

Early 100s had some problems with durability in the areas where the hitch mounted, but this was soon corrected.

Both the 100 and 200 used a slightly revised (and slightly different between them) C-123 engine. The 100 engines ran at 1,400 rpm, while the 200 engine ran at 1,650.

Farm Tractor Decision 1879 eliminated the International 100. Industrial tractor features could be ordered with the Farmall 100, those features included the foot accelerator, heavy-duty front axle, front lamp bracket, industrial tires, and the deduct for Touch Control hydraulics. However, some customers wanted a separate tractor. To meet demand, IH reinstated the I-100, which was actually a Farmall 100 with an International name attachment consisting of nameplate and hood.

The 130s were descendants of the 100s, which descended originally from the Farmall A. The 130 and 130HC received the increased horsepower C-123 engine that operated at 1,800 rpm instead of the 1,650 of older tractors. The crankcase, sleeves, valves, governor, and crankshaft bearings received some updates. A new manifold was used, and new carburetors from Zenith and Carter were used. A new IH distributor designed to give maximum advance at 1,800 rpm completed the engine.

The 130 chassis was updated to handle the new speeds and to increase part commonality with that of Farmall 230s. A new air cleaner; carburetor; rear lamp supports taken from the 300/400 Series tractors (again increasing commonality); and a new starting-switch pull-rod assembly were used. The paint scheme and white nameplate decal, along with a new serial-number plate reflecting the new model number, were the other major changes. Tractor speeds were increased due to the increased enginer pm.

The Farmall 230 was also a 200 with some more extensive revisions than its little brother. The increased horsepower C-123 was similar to the 130, but a new, single-valve Hydra Touch hydraulic system was used, as well as the new weight-transfer articulated-linkage Fast Hitch. A two-valve Hydra Touch system and a rear hydraulic manifold were available as special attachments. Weight transfer allowed the draft of the implement to be transferred as downward force to the rear traction wheels, providing additional traction, critical in small tractors. After the hydraulics, which introduced weight transfer into IH's small tractors for the first time, the major changes were paint, decals, and a new serial-number plate. A new toolbox and seat assembly eliminated the provisions for the old hydraulic hose tunnel, rockshaft bearings, and other older parts no longer used on the new tractor. The higher-rpm engine increased tractor speeds.

Cotton Picker Farmall 200 and 230

Farmall 200 and 230 tractors were used as the base unit for the 2C-14 and 2CK-14 cotton pickers manufactured at IH's Memphis, Tennessee, factory. These tractors had a large number of modifications from the standard tractors. However, packages were available to convert the tractors back to use as farm tractors when they were not

needed for picker duty. Included were exhaust pipes, a few different parts for the electrical system as appropriate, drawbars, and other parts.

The tractors were apparently shipped to Memphis without any modification. The tractor had to have hydraulic remote control or Hydra Touch equipment. For attachment 0359775R96 (which covered the 200 and 230) at Memphis, the belt pulleys and power takeoffs were removed and cover 9092-D was installed. Because the tractor would be operated in reverse with the picker attached, the dual front wheel bolster was rotated 180 degrees to provide the castor towards the front of the tractor. The drive bevel gear was to be assembled on the righthand side of the tractor, which would reverse the direction of travel. The tractors had to have the 88-inch rear-axle attachment, and it was recommended that the tractor be fitted with 5.00x15 4-ply front tires, and 10x36 6-ply rear tires.

For attachment 362097R93 on the 230, the pre-screener attachment 353410R92 was mandatory, along with the above requirements. The rear tires were installed backward from normal as well. A power takeoff was used with this attachment, and front-frame axle-support bearings were used, located between the axle carriers and the rear wheels of the tractor.

The attachments consisted of a different exhaust system, complete with spark arrester, which directed the exhaust down so that it wouldn't set the cotton on fire or interfere with the operation of the picker. A plug was used to cover the regular exhaust hole. A "Danger Reverse Operation" (1004451R1) tag was affixed to the tractor. A variety of changes to the operator's position and controls were also fitted, including clutch and

The Farmall 100 was a Super with new sheet metal, a one-point Fast Hitch as an option, and other minor changes. As usual, a variety of options were available—this one has the heavy-duty square-tube front axle and a foot accelerator. It looks like this tractor's ignition is original, probably the same coil as new. The major difference between the International 100 and the Farmall is the nameplate. Tractors as nice as this are increasingly being preserved in original condition.

The 100 series tractors shared the same layout as the Farmall A, including the offset operator's position, known as "Cultivision." Note that this tractor has the foot accelerator and heavy-duty square-tube front end, features most often associated with the International-type 100s. In fact, the Farmalls could have the same features (and did in the previous tractor), while later Internationals did not necessarily have to have those features after the International stopped being a separate model.

brake pedals and electrical systems (362098R91 only), to name a few.

Kinds and Codes
Sales packages were available for all these tractors that included the basic tractor and selected options. The tractors sold as parts of these packages probably comprise most of what collectors see surviving today.
Farmall 100 Domestic Kind 503 Codes 1, 2, 3 (Code 2 added November 24, 1955, Code 3 added December 5, 1955)
Farmall 100 HC Domestic Kind 505 Code 1
Farmall 100 Export Kind 503 Codes 1, 2, 3 (Code 3 added February 17, 1956)
Farmall 100 HC Export Kind 505 Code 1
Farmall 130 Domestic Kind 806 Code 1
Farmall 130 HC Domestic Kind 806 Code 2
Farmall 130 Export Kind 806 Code 1
Farmall 130 HC Export Kind 806 Code 2

Engine
The 100 and 200 tractors had a C-123 engine similar to that used in the Super C and the Super A-1. However, the 130 and 230 tractors had C-123 engines with a variety of upgrades (new crankcases, valves, cylinder head, sleeves, manifold, carburetor, piston rings, and related parts).

Gas
The C-123 gasoline-burning engine was standard in the 100, 130, 200, and 230. The "C" stood for "carbureted," while the "123" referred to the engine displacement. High-altitude pistons were also available. The 230 offered 12 percent more horsepower than the 200.

C-123 engines starting with serial 501 were used on the 100 and 200; C-123 serial 36001 was the starting engine for the 130 and 230. The later 140 and 240 models used engines beginning with serial number 65000. Further changes occurred on the 140 and 240 engines, which are beyond the scope of this book, although those engines might show up as replacement engines. Engines were numbered sequentially as they left the engine assembly line, without regard for where the engine was to be applied. The C-123 was used in a variety of other products, including the U-123 power unit and engine, the 76 Harvester-Thresher, and the 55-Series hay balers.

Kerosene/Distillate Options
Distillate and kerosene engine versions were available but were rarely ordered options (foreign customers mainly chose them). The equipment was offered either from the factory or as a field attachment. The kerosene and distillate engine options included radiator shutters to help maintain the high temperature needed to keep the heavy fuels vaporized. All attachments had the small gasoline-starting tank, with the appropriate hood with the opening. The fuel strainer was different to accommodate two fuel inlets. A hood heat shield was fitted to prevent heat from the manifold from affecting the hood (367734R1). Exhaust pipe 356369R1 was used. An oil pan with drain plugs was used so that unburned fuel could be drained each day. Two different drain plugs were used in production.

Specialized Zenith 67x7 or Carter UT carburetors were used on the distillate—and kerosene-burning tractors.

Also included in the attachments were decals describing how to adjust the fuels during the starting process. Decal number 1000678R2 was the English version, 1000769R2 was Spanish, and 1000821R2 was French.

Engine Suffix Codes
Originally, suffix codes were A (distillate), B (kerosene), D (5,000-foot altitude), and E (8,000-

Table 3.1
Farmall 100 and 130 Kinds and Codes

Attachment	Kind 503 Code1 Dom	Kind 503 Code 2 Dom	Kind 503 Code 3 Dom	Kind 503 Code 1 Exprt	Kind 503 Code 2 Exprt	Kind 503 Code 3 Exprt	Kind 505 Code 1 Dom	Kind 505 Code 1 Exprt	Kind 806 Code 1 Dom	Kind 806 Code 1 Exprt	Kind 806 Code 2 Dom	Kind 806 Code 2 Exprt
356121R91 Pneumatic front 3.00s z 15 DC rim	x			x					x		x	
356122R91 Pneumatic Adjustable front wheel with 3.00D x 15 DC Rim		x	x		x	x						
353061R91 Front Wheel tire and tube 5.00x15 4 ply F-2 tread	x	x		x	x	x			x		x	
256274R91 Front Wheel tire and tube 5.00x15 4 ply F-3 tread			x									
356138R91 Rear tire and tube 9-24 4 ply R-3 tread			x									
356128R91 pneumatic rear wheel and W8-24 rim	x	x	x	x	x	x			x		x	
356111R91 Rear tire and tube-9-24 4 ply R-1 tread	x	x	Note 1	x	x	x			x		x	
359653R91 Deluxe upholstered seat		x	x		x							
364850R91 Foot Accelerator			x			x						
488831-DB Exhaust Muffler	x	x	x	x	x	x	x	x	x	x	x	x
363250R91 Fast Hitch	x	x		x	x							
70335-DC Heavy Duty Front Axle			x			x						
356576R93 Adjustable Front Axle	x	x		x	x							
353195R91 Power Take-off			x			x						
362629R91 Kerosene Burning Note 2			x	x	x	x		x			x	
063866-DA Oil Filter Attachment			x	x	x	x		x			x	
356126R91 adjustable front wheel and 3.00x19DC rim							x	x		x		x
356132R91 rear wheel and W8-36 rim							x	x		x		x
356145R91 rear tire and tube 9-36 4 ply R-1 tread							x	x		x		x
356136R91 Front tire and tube 4.00x19 4 ply F-2 tire							x	x		x		x
362631R92 Kerosene burning												x
364851R91 Touch Control used with AI-23 Mower			x			x						

Note 1–was required, eliminated April 16, 1956. Note 2–362623R91, distillate burning attachment, was replaced with 362629R91 kerosene burning attachment February 25, 1955. Kind 503 Code 3 was for Farmall 100s setup for AI-23 Hi-Way mowers.

Table 3.2
Farmall 200 and 230 Kind and Code Chart

Farmall 200 Domestic Kind 507 Code 1, 2, 3 (Code 3 added 11-24-1954)
Farmall 200 Export Kind 507 Code 1, 2, 3
Farmall 230 Domestic Kind 811 Code 1
Farmall 230 Export Kind 811 Code 1

Attachment	Kind 507 Code 1 Dom	Kind 507 Code 2 Dom	Kind 507 Code 3 Dom	Kind 507 Code 1 Exprt	Kind 507 Code 2 Exprt	Kind 507 Code 3 Exprt	Kind 811 Code 1 Dom	Kind 811 Code 1 Exprt
356121R91 Peeumatic front 3.00s z 15 DC rim	x	x		x	x		x	x
353061R91 Front Wheel tire and tube 5.00x15 4 ply F-2 tread	x	x	x	x	x	x	x	x
356170R92 rear wheel and W9-36 Rim	x	x	x	x	x	x	x	x
356146R91 Rear tire and tube 10-36 4 ply R-1 tread	x	x	x	x	x	x	x	x
48831-DB Exhaust Muffler	x	x	x	x	x	x	x	x
063866-DA Oil Filter element package				x	x	x	x	
362649R92 Kerosene Burning				x	x	x	x	
359570R96 Fast Hitch	x	x	x	x	x	x		
357933R92 Adjustable wide tread front axle		x			x			
357819R92 Deluxe upholstered seat		x			x			
356122R91 Pneumatic Adjustable front wheel with 3.00D x 15 DC Rim		x			x			

Kind 507 Domestic code 3 had the deluxe type seat added November 3, 1955. Kind 507 Export Code2 had the deluxe seat added October 6, 1955

The International was often sold through IH's industrial dealerships to municipal and commercial customers, most likely for roadside and grounds maintenance. Heavy-duty sickle mowers especially designed for these tractors were also sold by IH.

Table 3.3
Piston, Cylinder head, and Compression Ratio Chart, C-123 (100 and 200 Tractors)

	Piston	Cylinder Head	Compression Ratio	Manifold
Gasolene	356302R2	251172R2	6.00:1	251175R1
Gasolene, 5000 Foot Altitude	356618R2	251172R2	6.87:1	251175R1
Gasolene, 8000 Foot Altitude	356619R2	251172R2	7.48:1	251175R1
Distillate	356302R2	351649R2	5.02:1	351665R1
Distillate 5000 Foot Altitude	356618R2	351649R2	5.59:1	351665R1
Distillate 8000 Foot Altitude	356619R2	351649R2	5.95:1	351665R1
Kerosene	356899R2	351649R2	4.65:1	351665R1
Kerosene 5000 Foot Altitude	356916R2	351649R2	5.24:1	351665R1
Kerosene 8000 Foot Altitude	356917R2	351649R2	5.82:1	351665R1

Table 3.4
Piston, Cylinder head, and Compression Ratio Chart, C-123 (130 and 230)

	Piston	Cylinder Head	Compression Ratio	Gasket	Manifold
Gasoline	362867R1	366206R1	6.8:1	366300R1	366305R1
Gasoline High Altitude	363375R1	366206R1	7.89:1	366300R1	366305R1
Distillate	366669R1	351649R2	4.85:1	366300R1	351665R2
Distillate High Altitude	363375R1	351649R2	5.74:1	366300R1	351665R2
Kerosene	356899R2	351649R2	4.62:1	366300R1	351665R2
Kerosene High Altitude	356916R2	351649R2	7.89:1	366300R1	351665R2

foot altitude). On March 21, 1957, the D and E codes were replaced by a U to signify high altitude. On November 5,1957, the code for the exhaust-valve rotator attachments (V) was added, but to the chassis seriel numbers.

Valve Cover
There were two optional breathers.

Valve Rotators
Valve rotators were available for the C-123 engine. Tractors with valve rotators had a V suffix on the chassis serial number.

Cylinder Heads
Cylinder heads for the gasoline C-123 tractors changed at engine serial number 36001 and up, with the old number being 0251172R2 (355046R21, the parts number in the parts book, was actually an assembly used for repair only). The newer number was 366206R1. The kerosene/distillate head was 261295R2; exhaust valves also changed at this serial number.

Crankshafts
Tractors without electrical starting were fitted with starting cranks as standard, although tractors with starters could have cranks as an attachment. The 130 and the tricycle 230s shared starting cranks, but 230s with adjustable front axles had a longer starting crank; cotton picker 230s had another different starting crank.

Choke
Tractors shipped without starters were fitted with front choke rods. On 100, 130, 200, and 230 tractors without starters, the rear choke rods were eliminated.

Electrical
Battery electrical systems were 6-volt. Battery ammeter 360052R91 was used. Tractors could be ordered without a starting and lighting system,

with a hand crank used instead. (Hand crank wasn't "in place", it was clipped to a much different part of the tractor that the place it was used!). Magneto ignitions were available, with and without starting and lighting systems.

Battery Ignition Unit
Battery ignition was standard for these tractors.

Magneto (IH H-4)
The Farmall 100, 130, 200, and 230 were available with IH magneto attachments. Several different versions of the electrical system were used with magnetos. Systems were available without starting and lighting systems, for use with a starting crank (which was shipped with the tractor for domestic customers, but curiously was not shipped for export). Other magneto ignition tractors were fitted with starting and lighting systems. Later parts books list Wico magnetos, but these were not used in production or service in the time of production. A different ignition switch was used for tractors with magnetos.

Headlights and Rear Lights
Tractors shipped to Switzerland received a different headlight and rear light package. The attachment, 360551R91, was composed of older-type non-sealed light units that were legal in that country. All other tractors used sealed beams.

International 100s, Farmall 100, and Farmall 130s were fitted with the industrial-tread front tires and heavy-duty front axles; 100 and 130 high-crops were fitted with front lamp attachment 364849R91, which mounted the headlights on longer posts that angled up, compared to the regular Farmall 100 and 130 tractors, which had the lights mounted on posts that came out a short distance above the front axle.

Combination Rear Light and Taillight
All tractors had available the combination rear lamp and taillight (large white light, with a small red lamp insert inside) available as an attachment for factory or field application.

Battery
Auto-Lite or Globe Union Batteries were used on the 100 and 200 Series tractors. About the time that the 130 and 230s came out, the Globe-Unions were eliminated. Battery cables 350736/360054R91 (battery to ground) were rubber-coated SAE #2 wire, with minor variations in the die cast terminal from the four different suppliers of the part. All suppliers cast a "P" into the terminal. Number 350735/360553R91 (battery to cranking motor) was made of #1 battery cable, rubber-covered, SAE #2 wire, with an "N" cast into the die cast terminal. There were minor differences in production between the four suppliers.

The starting and lighting (battery Ignition) wiring harness, 354243R92 or 365058R91, was made with a woven covering impregnated with either asphalt or varnish. The second part numbers in each type of wiring set were removed April 27, 1955, when electrical equipment made with emergency substitute materials was eliminated.

Electrical Breakaway Connector and Safety-Lamp Package
Safety-lamp packages were available to mount on implements with the cord running to a breakaway socket mounted on the tractor. The 897104 lamps (I want to emphasize that the 897104 part number was a Guide Lamp number, which the sentence construction doesn't do.), made by Guide Lamp, were 6-volt. When the tractor was not fitted with fenders, the lamp was mounted onto the tractor with bracket 364290R11. The electrical breakaways were mounted on the lefthand seat

Tractors of this era had tires with lugs at a 45-degree angle from centerline, like these. In later years, a 23-degree angle was found to be more efficient, and nearly all modern tires have this chevron pattern. The older-style 45-degree tires can still be purchased and are being reproduced for collectors.

Chassis

Radiator, Hood, Grille, and Nameplates

Export tractors had different hood assemblies, drilled differently to handle the different nameplates. Farmall 130 and 230 tractors had identical hood and hook assemblies, with the 130 HC (high clearance) having a slightly different hood drilled for the different nameplate. The Farmall 100 and 200 had identical hood assemblies, but the Farmall 100 HC and the International 100 had different hood sheets, again drilled differently for nameplates.

The hood hooks were changed November 12, 1954. The new hood clamp assemblies, 363668R91 and 363669R91, were designed so that gasoline and distillate/kerosene-burning tractors could use the same hood clamp assemblies. Before this, the distillate/kerosene–burning tractors had to have different clamp assemblies on the right side (60754-D).

IH purchased the serial-number plate printed, and with the model prefix and number pre-stamped. Suffixes that depended on equipment were hand-stamped at the factory. The plates were changed March 2, 1955, to remove the maximum idle speed and overload warning. Originally, the International 100s and Farmall 100s had different serial plates (0362511R2 for the International, versus 362510R2 for the Farmall), but apparently the International serial plate was eliminated at the same time the separate International was eliminated. Either brass or steel 11 gauge x ⅜ inch-long escutcheon pins were used to fasten the plate.

The IH emblem was on a plate on the front of the hood. On the sides on domestic tractors were separate plates for "Farmall," "McCormick," and the model number. The latter three were stainless steel. There was an "International" nameplate for export and industrial tractors. The nameplates all changed (authorized December 20, 1954) to five brass, attaching studs, replacing three stainless-steel studs of shorter length. The tubular clips changed as well, and hoods received extra holes for attaching the nameplates.

Farmall 100 HCs also had a separate "Hi-Clear" nameplate, again of stainless steel.

Radiator, Shutter, Controls, and Heat Indicator

Two different radiator cores were authorized for production, 356356R94, produced by IH, and 358421R93, produced by Young. The radiators differed from Super A-1 and Super C produc-

The operator positions–the hydraulic control levers are visible. The shifter knob was painted red in production. The steering wheel was probably attached after painting. Blueprints for the wheels specify that the suppliers are to paint the spokes and center red.

bracket for the 100 and 130. On the 200, with the lamp on the left side of the tractor, the breakaway was located on the outside hole on the right side of the platform, or the outside mounting hole for the PTO cover, with the 6-volt warning decal located either on the belt-pulley guard or on the lower right hand side of the battery box. On 200s, with the lamp mounted on the right side of the tractor, the breakaway socket was mounted on top of the lamp bracket, with a decal located in the same area. When fitted with the electrical system, the tractor had an electrical caution decal 1001725R1. Two different sockets were used, one manufactured by Cole Hersee, the other by the Joseph Pollack Co.

The Farmall 200 was essentially a restyled Super C with some improvements. It was a small, two-row cultivating tractor that was very pleasant to operate and had a good reputation.

tion (except possibly late production) in having a filler neck designed for use with "lobed" radiator caps used with 100 and 200 tractors. The radiator sat on pads, 47240-D, which were made from either gulf-cloth inserted-sheet packing purchased from W.H. Salisbury & Co., or from tire carcasses purchased from A. Lakin and Sons Inc., Chicago.

Originally, radiator caps in production were 361705R91 (Stant) and 362323R91 (AC Spark Plug). On January 22, 1957, the specifications were changed. The general radiator cap number became 361705R91, with caps provided under this number by Stant, AC Spark Plug Co., and Eaton Mfg. Co.

Kerosene and distillate-burning tractors came equipped with a radiator shutter attachment. The shutter could be ordered with gasoline-burning tractors as well. The shutter was, like the kerosene and distillate equipment, available factory or field. The attachment contained the shutters, the control rods, and the lever (spring loaded) and sector at the operator's end. A decal describing the operation of the shutter was included (1001357R1 English, 1001358R1 Spanish, 10011359 French). A heat indicator was used on tractors to indicate when to operate the shutter (362347R91).

An attachment or parts-accessory package that supplied a different thermostat and heat indicator for use with low-boiling-point anti-freeze was available throughout production of the 200, 230, 100, and 130 family tractors. The packages included a thermostat that operated in the 130–150-degree range; the heat indicator was part number 351698R92.

The radiator-hose clamps were made by Wittek; the inlet clamp was Wittek GM-55 (IH Part number 29976-D), and the outlet clamp was Wittek GM-43 (IH part number 29970-D)

Governor and Choke Controls

Farmall 100 and 130 tractors could be fitted with a foot-accelerator attachment at the factory. The International 130 was fitted with the foot accelerator as standard. The actual lever was made of red brass (9151-D).

35

This tractor has a mounted two-row planter. In this era, tractor-mounted implements—especially front-mounted implements–are designed to mount to a specific model of tractor. Some, including planters, can be quite rare today. Fast-Hitch implements are more standardized.

Fuel Tanks, Strainers, and Caps

Fuel-tank cap 23995-DC was standard on all 100, 130, 200, and 230 tractors on the main fuel tank but should be refitted today with the "safety" gas cap. Beginning February 6, 1956, a fuel-gauge filler cap (365481R91) was authorized for factory installation as an attachment for the Farmall 100 and 200. It was available for Farmall 130 and Farmall 230 tractors throughout production.

The front, fuel-tank support bracket had two different parts that were optional: 356008R21 (sheet steel) or 356495R1 (gray iron).

Air Cleaners

Air cleaners 366320R1 (Donaldson) or 367765R91 (United Specialties) were used with the Farmall 130 and 230 Series. Air cleaners 351187R92 (Donaldson) or 351188R91 (United Specialties) were used on all versions of the 100 and 200 Series. There were minor differences between the cleaners, with some having a date stamped on the shell. All had the instruction decal in the appropriate language of English, Spanish, or French.

Clutches

Auburn and Rockford clutches were used.

Clutch Housing

Clutch housing covers and springs were the same on all 100, 130, 200, and 230 Series tractors, and a piece of felt went between the cover and the housing. Two different clutch-housing covers were used optionally with each other, 8778-DA (gray iron) or 354909R1 (sheet steel). The hand hole cover was also the same between the tractors. The 230 had a different clutch housing from the 100, 130, and 200 tractors. Clutch housing 366615R11 was apparently used in production on early Farmall 130 and 130 HC tractors, but in January 1957, clutch housing (basic housing was 351687R8), which was used previously on Farmall 100 tractors, was reinstated as the production housing; 366616R11 was still available for service use. The reason for the change was to reduce cost.

Front Axles

All 100 and 130 Series tractors used a wide front axle, although several different versions were available. A fixed non-adjustable front axle was available for the Farmall 100.

Adjustable front axles were the standard front axle for the Farmall 100. The 230 also had an adjustable wide-tread front axle as a no-charge option. The axles could be added as a field option if the tractors were not already fitted with them. Treads of different widths were obtainable in 4-inch increments with the axle, with additional adjustments possible with adjustable wheels. On the 200, the minimum tread (with 4.00x15 tires, adjustable wheels) was 50 inches, while the maximum tread was 83¼ inches. On the 230, the minimum tread was 56 inches, while maximum became 89¼. The wide front axles were available with either adjustable or non-adjustable front wheels.

Heavy-duty adjustable front axles (square tubes, as opposed to round tubes on the other axles) were available as optional equipment on the Farmall 100 and 130 as a factory attachment, 70335-DC. This was the same axle used on the International 100.

The 200 and 230 Series tractors could use a single front wheel, a two-wheel narrow front wheel, or a wide front axle, with varying versions of the wide axle available. A double, narrow front wheel and bolster attachment was available (either as an attachment or a parts accessory) in the field for 200 and 230 tractors that were shipped with either a wide front axle or a single-wheel narrow front. This attachment was available from November 7, 1955; 5.00x15 4-ply tires were used with this attachment.

Front Wheels

All wheels and tires were officially considered to be options and were to be ordered separately. However, IH did establish a default to ship when salesmen forgot to order a wheel/tire combination. On the 200 and 230, the default was the 356121R91 attachment with the 3.00x15DC rim, with 5.00x15 4-ply F-2 tread tires.

On the Farmall and International 100 and Farmall 130, part 356121R91 was the front wheel with the 3.00Dx15DC rims, while the Farmall 100 HC and 130 HC got 356126R91, including 3.00Dx19DC rims. On the Farmall 100, International 100, and the Farmall 130, the 5.00x15 4-ply F-2 tire was used; the Farmall 100 and 130 HCs got the 4.00x19 4-ply F-2 tire.

Dual Front Wheels

A dual-wheel front end was one of two non-extra-cost front ends for the 230. The dual-wheel configuration was available with either fixed or

Table 3.5
Front Rim & Adjustabe Front Wheel Attachments 100 and 130

Attachment number	Rim Size	Rim Manufacturer	Tire Size	Plies	Tread
356121R91	3.00Dx15DC	NA	NA	NA	NA
356122R91	3.00Dx15DC	Cleveland Welding	4.00X15 Note 1	4	F2
		Goodyear	5.00x15 Note 1		F2
		Firestone	4.00x15 Note 2		F-3
		Electric Wheel	5.00x15 Note 2		I-1
356123R91	4.00Ex16DC	350850R91	6.00x 16 Note 2	4	F2
356124R91	4.50Ex16DC	392760R91	6.00x 16 Note 3	4	F-2

Note 1 used with adjustable front -axle attachment 356576R93
Note 2 used with heavy- duty front -axle attachment 70335-DC and front lamp bracket attachment 364849R91
Note 3 Used with adjustable front- axle 356576R93 OR heavy-duty Front -Axle Attachment 70335-DC

Table 3.6
Front Tire and Tube Attachments

Attachment number	Tire Size	Plies	Tread	Used with Rims
356106R91 Note 1	4.00x15	4	F-2	3.00Dx15DC
356135R91 Note 2	4.00x15	4	F-3	3.00Dx15DC
353061R91 Note 1	5.00x15	4	F-2	3.00Dx15DC
368100R91	5.00x15	4	F-1	3.00Dx15DC
254763R91 Note 4	5.50x16	4	F-3	4.00Ex16
354108R91 Note 1	6.00x 16	4	F-2	4.00Mx16DC
365346R91 Note 3	5.50x16	4	I-1	4.00Ex16DC

Note 1 Farmall 100 originally, not International. Note 2 Originally International 100 only, then added to Farmall 100 when the International nameplate was authorized for use on the Farmall September 15,1955. Note 3 Added December 12, 1955. Note 4 Eliminated December 12, 1955. Note 5 23705-S replaced 062412-D 12-12-1955 (Tire manufacturers eliminated 5.50-16 tires in F-3 treads due to low demand).

Table 3.7
130 HC Front Rim & Disc Attachments

Attachment number	Rim Size	Rim Manufacturer	Tire Size	Plies	Tread
356125R91	4.50E x 16DC	392760R91	6.00 x 16	4	F-2
356126R91	3.00 D x 19DC	Goodyear	4.00x 19	4	F-2
		Firestone			
		Cleveland Welding			

Table 3.8
100 and 130 HC Front Tires and Tubes

Attachment #	Tire Size and Tread	Plies	Tread	Used with Rims
354108R91	6.00x16	4	F-2	4.00Mx16DC
356136R91	4.0019	4	F-2	3.00Dx19DC

Table 3.9
200 & 230 Front Rim & Disc Attachments

Attachment number	Rim Size	Rim Manufacturer	Tire Size	Plies	Tread
356121R91	3.00x15 DC	?	?	?	?
356122R91	3.00Dx15DC	57095-D or 350327R91	4.00 x 15	4	F-2
			5.00 x 15	4	F-2
356126R91	3.00Dx19DC	GET	4.00 x 19	4	F-2
356166R91	4.50E x16	GET	6.50 x 16	4	F-2

Hub caps were either gray iron (part 365144R1) or sheet steel (49115-D) until April 29, 1955, when the gray iron cap was discontinued.

Table 3.10
200 and 230 Front Tires and Tubes

Attachment number	Tire Size and Tread	Plies	Tread	Used with Rims
356106R91	4.00x15	4	F-2	3.00D x 15dc
353061R91	5.00x 15	4	F-2	3.00x 15DC
354784R91 Note 1	6.50x16	4	F-2	4.50E x 16DC
355748R91 Note 2	7.50x10	6	F-2	Single Front
356136R91	4.00x19	4	F-2	3.00 x 19DC
361102R91 Note 2	6.00x12	6	F-2	Single Front

Note 1 For Adjustable wide front axle. Note 2 For Single Front Wheel.

A view of the instrument panel with what looks like all-original gauges. The black knob under the push/pull ignition switch is the fuse.

variable-tread wheels. The variable-tread front wheels could be set to 6¾, 9 ?, or 12¾-inch treads, and was extra equipment.

Single Front Wheel

A single front wheel was available as extra equipment (factory or field) on the 200 and 230. The wheel was available in 7.5 x 10 6-ply (which had a split rim and wheel) or 6.00x12 4-ply-pneumatic tires (solid rim and wheel). The single front wheel was used for working crops (mainly vegetable and root crops) with narrow row spacing. The axle was held on each end by the front wheel fork. The equipment was essentially the same as used on the C and Super C tractors.

Attachment #	Tire Size and Tread	Plies
57894-DA	7.5x10	6
58282-DA	6.0x12	4

Steering Mechanism

On the 230, an enclosed worm-gear-type steering unit was used. An 18-inch steering wheel was supplied by Sheller (60070-D) or French & Hecht (29118-DC), which, among other small differences, had slightly different spoke contours. On April 29, 1955, optional steering bracket 365002R11 and 0365002R1 (gray iron) were eliminated by FTC-12490 (elimination of substitute material parts) on the F-100 tractors.

Front Wheels and Rims

Standard front wheels on the Farmall 100 were non-adjustable. Adjustable front wheels were available as extra equipment.

Gear Shifter Mechanism and Transmission

A seven-mile-per-hour fast tillage speed was available as extra equipment on the 200 and 230. This speed replaced the normal third-gear driven gear with a 30-tooth gear (359978R1) and third- and fourth-gear sliding gear (the new gear was 359979R1, with 31 and 37 teeth). All tractors used the same gearshift assembly and the same gearshift ball, 53717-DA. Early 100 family tractors had two different transmission rear covers authorized: one of sheet steel (46060-D) and one of gray iron (365143R1). The gray iron cover was eliminated April 29, 1955, when parts using substitute materials for emergency use were eliminated.

Final Drive and Rear Axles

All 100 tractors originally had specified rear axle housing (lefthand) 362771R11 for production, but in late October 1954, that housing was changed to 100 HC only, and the International and Farmall 100 switched to 0363273R91 for production, while the original housing was still used for service. Housing 363273R91 was made of gray iron. The same situation occurred with the right hand housing, where 362772R11 was the original housing, and 0363274R91 was the new housing for use on the Farmall and International 100.

The 200 had a straight axle where the wheels slipped in and out on the axle to adjust wheel tread. Originally on the 200, two different rear axles were optional with each other, 365147R2 and 351115R3. These axles were changed by a decision dated April 15, 1955, which changed the method of fastening the bull-gear assembly to the axle to facilitate disassembling the bull gear for export shipments. Previously, a bull-gear retainer and cap screws were used; the new method was a snap ring.

The 365147R2 axle was eliminated April 29, 1955 with many other emergency substitute parts.

Brakes

Double disc brakes were standard. A field change was available for Farmall 200s below serial number 11118 to update the actuating disc. This package was voided June 27, 1957.

Rear Wheels and Rims

IH considered all rear wheels and rims to be attachments that were to be ordered separately. If the salesman forgot to specify the equipment, IH shipped a default package.

On the 200 and 230, the rim and rear wheel default was 356170R91, which included the W9-36 rim attachment. The tire default was 356146R91, which included the 10x36 4-ply R-1 tread rear tire.

Seat and Platform

The standard seat on the Farmall 100 and 230 was a spring-mounted non-upholstered pan seat, 363488R91, made by Milsco. Upholstering was available as extra equipment. A back seat flip bracket was available on the 200 and 230 for factory or field attachment. Seat springs for the 200 and 230 were available in light (354673R92), medium (360083R91), and heavy versions (354672R91) to accommodate operator weight. A clip mounted to the seat or bracket was used with trip ropes for clutch lift implements.

Seat locator knob 352659R1 was used, made of sheet steel by Monroe.

Deluxe Cushion Seat

Farmall 100 and 130 Series (including HC) had the deluxe cushion seat attachment available factory or field as of May 1956. Essentially, the same as the 300 utility seat; the covering was in cotton duck, coated with silver Koroseal. These were square seats with backs and tube arms.

Armrest Pad

For Farmall 100, 130, 100 HC, and Farmall 130 HC tractors with the deluxe cushion seat attachment, an armrest attachment was also available factory or field. There were actually two different pads available under part number 363740R91, from Milsco Mfg. Co. or Bostrom Mfg. Co. Both were covered with silver Koroseal and were continued after 1958 with imitation black leather.

Deluxe Type Upholstered Seat Attachment

This seat attachment was available on Farmall 100, 100 HC, 130, 130 HC, International 100, Farmall 200, and 230 for factory installation throughout production, and field application on most tractors. This essentially was an upholstered pan-type seat, covered with cotton duck coated with silver Koroseal.

Also available were the silver Koroseal-type detachable seat pads that could replace older seats after the original was cut out with a knife, and were available for the older tractors produced after 1939.

Detachable Seat Pad

Detachable seat pads available for tractors with the plain pan seat, 351438R92, were handled as the deluxe type; 359483R91 were the plain type. The seats were covered with silver Koroseal-coated duck, and had molded bottoms to fit the metal pan.

Table 3.11
100 and 130 Rear Wheel and Rim Attachments

Attachment number	Rim Size	Rim Manufacturer	Tire Size	Plies	Tread
356128R91	W8-24	Firestone	9-24	4	R1,R-3
		354673?R92	9-24	4	R-3
		Goodyear	8-24	4	R-3
		Cleveland Welding	11-24 See Note 1	4	R-1, R-3 See Note 2
		Electric Wheel See Note 3 51261DA			
356129R91	W-9-24	Firestone	10-24	4	R-1
		Goodyear	10-24	4	R-1
		Cleveland Welding	10-24	4	R-3
		Electric Wheel 354824R92 56205DA	11-24	4	R-1

Note 1 deleted February 4, 1957. Note 2- R-1 tread added June 18, 1957. Note 3 Electric Wheel added August 26, 1955

Table 3.12
100 and 130 Rear Tires and Tubes

Attachment #	Tire Size and Tread	Plies	Tread	Used with Rims
356111R91	9-24	4	R-1	W8-24
356137R91	8-24	4	R-1	W8-24
356138R91	9-24	4	R-3	W8-24
356139R91	9-24	4	R-3	W8-24
356141R91	10-24	4	R-1	W9-24
356142R91	10-24	4	R-3	W9-24
356143R91	11-24	4	R-1	W9-24
368954R91 See note 1	11-24	6	R-3	W9-24

Note 1 Added June 18, 1957

Table 3.13
100 and 130 HC rear wheel and rim attachments

Attachment number	Rim Size	Rim Manufacturer	Tire Size	Plies	Tread
356132R91	W8-36	64983-D	9-36	4	R-1
		Goodyear Firestone Cleveland Welding Electric Wheel	10-36	4	R-1
356134R91	W10-36	352348R91	11-36	4	R1 See Note 1

Note 1-Jacksonville Florida branch only

Table 3.14
100 and 130 HC Rear Tires and Tubes

Attachment number	Tire Size and Tread	Plies	Tread	Used with Rims
356145R91	9-36	4	R-1	W8-36
356146R91	10-36	4	R-1	W8-36
356147R91 Note 1	11-36	4	R-1	W10-36

Note 1-For Jacksonville Florida branch only.

Fixed Drawbar

The standard drawbar on all tractors was a fixed drawbar. On the 100 and 130 tractors, this was the quick-attachable type. The high clearance –type was of course different than the regular due to the different dimensions.

Instrument Panel

The 100 and 130 tractors used an instrument panel that was located in an oval-shaped can. The steering wheel support was located in the right

Table 3.15
200 and 230 Rear Wheel and Rim Attachments

Attachment number	Rim Size	Rim Manufacturer	Tire Size	Plies	Tread
356169R91 Note 1	W8-36	Electric Wheel Firestone Goodyear Cleveland Welding	9-36	4	R-1
356170R92	W9-36	Firestone Electric Wheel	10-36 11-36 Note 2 9-36 Note 2	4 + 6 6 4	R-1 R-2 R-1
357496R91 Note 3	DW8-42	Unknown	9-42	6	R-1
369009R91 Note 4	W9-36 Heavy Duty	Electric Wheel	10-36 11-36 9-36	4 + 6 6 4	R-1 R-2 R-1

Note 1 Eliminated October 26, 1955. Note 2 Added May 23, 1956. Note 3 Added October 26, 1955. Note 4 Added June 28, 1958

Table 3.16
200 and 230 Rear tires and Tubes

Attachment number	Tire Size and Tread	Plies	Tread	Used with Rims
356145R91	9-36	4	R-1	W9-36
356146R91	10-36	4	R-1	W9-36
356174R91	11-36	6	R-2	W9-36
357499R91	9-42	6	R-1	DW8-42
359487R91	10-36	6	R-1	W9-36

side of the can. The whole assembly mounted on the left side of the tractor

The 200 and 230 Series tractors have a small instrument panel located in the steering wheel support, in front of the shifter lever. The panel held the charge indicator, the ignition switch, and the lighting switch from left to right on the top, while the heat indicator was the only gauge on the bottom point of the panel. The oil-pressure gauge was still located up by the oil filter; on the 200, a Touch Control fluid temperature gauge was located above the hydraulic levers.

Hydraulics

The tractors had live hydraulic pumps, produced by Pesco Products Co. and by Thompson Products Co., driven from the camshaft gear.

Touch Control was standard on the Farmall 100, 130, and 200 Series tractors. Remote control for hydraulic cylinders on implements was optional and could be installed either at the factory or in the field. Fast Hitch was a factory or field attachment on the later Farmall 100 and all Farmall 130s, but was not available on the high-clearance tractors. Fast Hitch could be added on Farmall 100s below serial number 1377, but only when a new operator's platform was added.

The 230 had the Traction Control with Fast Hitch for direct-mounted Fast Hitch implements; and had the Hydra Touch hydraulics (a factory or field attachment) for implements with remote hydraulic cylinders, as well as the tractor-mounted hydraulics. A "Pilotguide" was mounted on the tractor to show the operation of the Traction Control of the Fast Hitch. The left hand scale showed the length the hydraulic cylinder was extended, while the right hand scale showed the weight transfer action.

The 100, 130, 200, and the 230 Series tractors could be ordered without the hydraulics (a "deduction" package). The hydraulic equipment was left off the tractor, but, in addition, a different fuel-tank bracket was used on the 100, 130, and 200 (49601-D), and the wiring harness varied slightly (an additional clip was used). The governor-control rod clip 350614R1 was also used. For tractors shipped without hydraulics, a hydraulic field package was available.

The 200-hydraulic deduction required a different battery box, a cover on the steering shaft support bracket (351364R1), as well as the front clip for the choke rod (352147R1).

The 100, 130, and 200 tractors had a field package available to add a hydraulic-lift heat indicator. The heat indicator was standard until July 1955, but was changed to a field attachment only on 100 and 200 tractors, IH having declared it "Not necessary for proper function of Touch Control System." The package was the same that was used on the Super A and Super C family tractors (and used the same booklet for installation instructions). The indicator was installed in the reservoir strainer. The wound metallic tube had two clips next to the bottom of the left side of the battery box, a clip near the top of the rear lefthand corner on the steering pedestal support, then the gauge clipped to the crosspiece of the hydraulic lever quadrant. The gauge was marked with "Touch Control Temperature," with two sections marked "Run" and "Hot," and had the IH logo at the bottom of the face. The gauge was carried under internal part number 356466R91 (Rochester) or 355841R91 (Stewart Warner).

All 100 and 130 Series tractors required a slightly different hydraulic package for use with the AI-23 Hi-Way mower. A conversion package was available to convert a standard Touch Control system for use with the AI-23.

Tractors fitted with Touch Control hydraulics received two different decals: the "Hydraulic Lift Oil Level" (1001348R1 English, 1001349R1 Spanish, 1001350R1 French), and the "Hydraulic

Lift Rockshaft Caution" (1001327R1 English, 1001351R1 Spanish, 100352R1 French).

Farmall 200 tractors used with the C-32 power loader needed a remote-control valve package that included reworking the head lamp support pipe, extra hydraulic piping, and valve assembly 356783R92 (body part number was 356782R2).

Remote-control packages in the 1950s included the valves, hoses to the breakaway couplings, the couplings, then the hoses and the ram itself. Tractors fitted with hydraulic remote control had one of three different hoses, all carried under part 356814R91. Manufacturers included Anchor Coupling, Flex-O-Tube, and Bestman.

Farmall 200s could be equipped at the factory with an attachment that would allow four arms at the rockshaft instead of the standard three. The 230 had an attachment available to convert a single-valve Hydra Touch system into a double-valve system. The attachment was available from the factory or as a parts accessory.

Farmall 200 and 230s fitted with Fast Hitch were required to have front-end weights that mounted on either side of the front bolster. They came in righthand (361159R1) and lefthand (361158R1) weights, each weighing about 86 pounds.

Attachments

International Nameplate

After the separate International 100 was discontinued, some customers still wanted an "International" tractor, so an attachment was authorized February 27, 1956, for those customers who wanted a tractor with that name. The attachment included two International nameplates with a hood assembly that was drilled to fit the nameplates, and was continued through Farmall 130 production into 1958.

Rear Axle Attachments

The Farmall 200 and 230 tractors had two optional rear axle widths to accommodate wider wheel settings for specialty crops. The standard axle width was 80 inches. The 88-inch wide attachment was available factory or field throughout production, with two different axle parts being used alternately (351479R2 vs. 365148R2); one was an emergency material substitution axle eliminated in May 1955. A 100-inch rear-axle attachment was also available throughout production. Besides the axle, the 100-inch rear-axle attachment included different rear axle carriers, and a different rear frame.

Air Cleaner Extension Pipe

An extension pipe was available to move the air intake up out of dust, especially in operations like haying or combining, or in dusty locations. The pipe was available on all 100, 130, 200, and 230 family tractors, factory or field.

Pre-Cleaners

Pre-cleaners were available as an attachment, factory or field—factory-only after June 2, 1955; afterward field availability was by parts accessory. Two different kinds that could be ordered: the collector-type made by Donaldson (362997R91) and the removable sleeve-type made by United Specialties (356249R92).

Pre-Screener

The Donaldson pre-screener was available from the beginning of production. On November 20,

The hydraulic levers in this view control the rockarms on each side of the tractor—the left lever controls the left side, and the right lever controls the right side. This can come in very handy for precision cultivating or planting, especially on hills or uneven ground. This view shows the hydraulic fluid temperature gauge, which can be quite useful in case part of the control system hangs up. Been there, done that!

A view of the rear hydraulic valve to the left and the starter to the right. In production, the starters were painted red along with the rest of the tractor. However, many replacement starters purchased from outside IH dealerships came in black. It's obvious from the condition of this starter's exterior paint that it hasn't been on this tractor its entire life.

1956, the United Specialties pre-screener was added, carried under a different parts number. On February 20, 1957, the two different pre-screeners were carried under the Donaldson part number. Up through June 3, 1955, the attachment could be ordered factory or field, but after that date field availability became a parts accessory.

Belt Pulley and Power Takeoff

The 200 and 230 used a combined belt pulley and PTO unit mounted on the rear of the tractor. Tractors fitted with the PTO got warning decals (1000636R5 English, 1000824R1 French, 1000772R5 Spanish). Many pulleys were purchased from Browning Mfg. Most tractors used an 8-inch diameter by 6-inch face pulley, with 6½ x 6½ and 7½ x 6½ pulleys by Browning available, and with a 7½ x 6½ gray iron pulley available from IH, as well. Some early 100s and 200s may have had Rockwood pulleys, but by August 18, 1955, IH canceled those (the paper pulleys); Rockwood pulleys had gone out of production at some time previous to that decision.

The 100 and 130 tractors had a rear PTO attachment; a combination rear power takeoff attachment and belt pulley; an angle rear power takeoff; or a combination rear and angle power takeoff. The 8½-inch diameter by 6-inch face pulleys were standard with the attachment, but 10½-inch diameter by 6½-inch face pulleys were available.

PTO shields differed slightly on tractors with belt pulleys versus tractors without belt pulleys; tractors with belt pulleys had a shorter shield because it was mounted on the longer belt-pulley drive housing.

Exhaust Muffler

The "standard" exhaust on all gasoline-burning tractors was steel exhaust pipe 356354R1. Kerosene- and distillate-burning tractors had a fitted pipe, 48892-DB, that was shaped to match the different exhaust outlet.

The exhaust muffler was officially an attachment, factory or field, on all 100, 130, 200, and 230 family tractors, although the muffler attachment was included on all the sales attachment packages. Originally, Mac Kenzie (48832DB), Hayes Industries (356004R91), Maremont Auto Products (357595R91), or Donaldson (360720R91) mufflers were used. All mufflers were made of aluminized steel, and had the IH logo and part number stamped into the body. On May 4, 1955, the Donaldson, Mac Kenzie, and Hayes mufflers were changed to a single part number, 360720R91; Maremonts were eliminated.

Spark Arrester

Spark arresters were available as factory or field attachments for all 100, 130, 200, and 230 tractors, throughout production. They were mandatory on cotton picker tractors.

Pneumatic Tire Pump
Both Schrader and Engineair (G.H. Meiser & Co.) spark plug–type pneumatic tire pumps were available as factory or field attachments for these tractors.

Swinging Drawbars
The Farmall 100, 100 HC, International 100, Farmall 130, and 130 HC all had available swinging drawbar attachments. Farmall 200 and 230s with Fast Hitch could be equipped with a swinging drawbar, factory or field.

Drawbar Pintle Hook
Farmall 100, 130, and International 100 tractors had available as an attachment a pintle hook drawbar. The pintle was available factory or field.

Fast Hitch
A Fast Hitch was not available on high-clearance models.

A one-point Fast Hitch was available on the 100 and 130 family tractors. An auxiliary drawbar was available for field installation, for use with PTO-operated machines.

The plain Fast Hitch was available on the Farmall 200, while Fast Hitch with Traction Control was available on the 230. A drawbar-stabilizer parts accessory package was available for field application on 230s with Fast Hitch. Farmall 200s and 230s ordered without a Fast Hitch could have one added as a field attachment. The 230s that had Fast Hitches added in the field also had to have the pilot-guide attachment added as well.

Auxiliary Drawbar
An auxiliary draw bar was for tractors equipped with Fast Hitch.

A small drawbar was available for Farmall 200 and 230 tractors with two-point Fast Hitch to plug into the two receptacles, it featured a flat bar between the point with the hitch holes in it. A drawbar extension plate could be fitted factory or field.

Cushion Spring
Farmall 100 and 130 tractors equipped with Fast Hitch could be fitted with a cushion spring attachment either at the factory or in the field. The spring absorbed shock from tillage equipment, especially in rough terrain and extreme subsoil conditions. In August 1955, factory

This tractor's wiring has been painted, leading the author to wonder whether the tractor has been repainted.

application was made available (previously it had been field-only). In September 1956, the field attachment became a parts accessory.

Hydra Creeper
Hydra Creeper units were available on Farmall 100, 130, 100 HC, 130 HC, 200, and 230 tractors as field attachments. Hydraulic lines from the regular pump supplied fluid to run a hydraulic motor in the creeper unit, mounted on the PTO shaft. The PTO shaft was turned, supplying power to the transmission. The clutch was disengaged when the creeper was activated through an interlock linkage between the creeper lever and the clutch. The gear ratio that resulted from the slower hydraulic motor (versus engine rpm) was about $\frac{1}{10}$ of the speed normal to the gear. The Hydra Creeper unit could be used in each gear, but no additional torque was gained; the tractor would stall out in the same respective gear with the same load if the clutch was engaged. The field attachments varied slightly based on the equipment of the tractor (Fast Hitch and hydraulic remote-control tractors got slightly different versions). Tractors fitted with the Hydra Creeper got Mounting Warning Decals (1001808R1 English, 1001809 Spanish, 1001810R1 French). The decals were authorized in July 1955.

Tachometer
Tachometers were available for the 130 Series tractors (370192R91). High-clearance units had a separate attachment (370193R91). The tachome-

The Farmall 130 was built in much smaller quantities than its Farmall-A and Super-A predecessors. It was essentially a restyled Farmall 100. Note that on the "McCormick" nameplate the space on the letters is painted white to match the white nameplate decal. Wisconsin Historical Society photo.

ter units were the only difference (370198R92 regular clearance, 370199R92 Hi-Clear).

Rear Wheel Weights (Non-Hi-Clear)
On 100 and 130 tractors, wheel weight 6701-DA was used, factory or field, with up to two on each side; each weight weighed 145 pounds.

The 100 and 130 Hi-Clears also used 145-pound weights, numbered 6818-DA (6818-D was used up until March 10, 1955). Also used either one or two to a side, a third weight could be added for better stability on the high-clearance vehicles by taking two bolts that attached the first weight to the second weight and, instead, used those bolts to attach the third weight to the second weight.

On 200 tractors, weights were available factory or field that were made of gray iron. Each weighed 140 pounds and were carried under part number 6818-D. Up to three weights could be put on each wheel.

Rear Wheel Weights (Hi-Clear) Split
On the Farmall 230 and Farmall 130 HC, split-type wheel weights were available factory or field. Both sides of the split had to be applied, with up to three sets applicable to each side of the tractor. Each half used part number 366109R1, and weighed 75 pounds. These weights were available for field application on 100 HC and 200 tractors, as well as Farmall AV, Super AV, Super AV-1, and the Farmall C and Super C.

Front Wheel Weights
The 100 and 130 (including Hi-Clear) and 200 and 230 tractors used wheel weights 6788-D, which weighed 42 ? pounds each. The weights could be applied two to a side.

Rear Wheel Fenders
Fenders were still optional on the 200 and 230, factory or field. The traditional "clamshell" fenders were used. The fenders were supported by U-bolts.

Seed-Plate Drive
A seed-plate drive was available factory or field on the 100 and 130 family tractors for use with side-mounted planters and dressers.

Cane Wagon Drawbar
The Farmall 100 HC and 130 HC had a quick-attach drawbar that was used only with sugarcane hauling carts. The attachment was available factory or field throughout production.

Decals
The 100, 130, 200, and 230 family tractors used very similar decals.

1000867R1 lighting switch
1000984R2 warning, pressure cooling (English, 1000985R2 Spanish, 1000986R2 French)
1000704R4 warning, brake (English, 1000767R4 Spanish, 1000819R3 French)
1000987R2 instruction, cylinder head breather (English, 1000988R1 Spanish, 1000989R1 French)
1001181R3 instructions, air cleaner (Donaldson), (English, 1001205R3 Spanish, 1001206R3 French) or,
1001289R1 instructions, air cleaner (United Specialities), (English, 1001290R1 Spanish, 1001291R1 French)
1001604R1 patent marking
1001182R1 warning, drawbar, and front pull hook (English, 1001184R1 Spanish, 1001185R1 French)
1001327R1 caution, hydraulic lift (English, 1001351R1 Spanish, 1001352R1 French)
1001348R1 hydraulic lift oil level (English, 1001349R1 French, 1001350R1 Spanish)
1001012R5 instructions, oil filter (English, 1001020R5 Spanish, 1001102R5 French)
1001718R1 Made in the United States of America (one required for export including Mexico, Canada, Puerto Rico, and Cuba), added October 22,1954; located next to serial number plate until February 22, 1955
2750013R1 background hood sheet nameplate, lefthand
2750014R1 background, hood sheet nameplate, righthand
1000636R5 power takeoff warning (English, 1000772R5 Spanish, 1000824R4 French)

The patent decals were changed in October 1956 to reflect additions or deletions to patent numbers; the part number received a step up in the R number.

The Farmall 230 was a revamped Farmall 200 with the new two-tone color scheme and a weight-transfer Fast Hitch. This example has a McCormick loader (along with a grill guard) that would also rate as a valuable collector's piece today. The 230 has a reputation as a very pleasant little tractor. Wisconsin Historical Society photo.

Chapter Four
Farmall 300, Farmall 350, International 300, and International 350

The Farmall 300 saw extensive changes from its Super-H predecessor. New were the optional Torque Amplifier, Independent Power Take Off, and Fast Hitch. Added with the new styling, the Farmall 300 was a full-featured row-crop workhorse for medium-sized farms.

The larger new, IH tractor lines saw more new features than the smaller tractors. The big news was the new 300 Utility. The "standard" tractor, the Super W-4, which was a traditional four-wheel plowing tractor, was replaced by a tractor based on the Ford 8N (in fact, an 8N was disassembled at Hinsdale during the engineering process, an industry-wide practice). This tractor was known throughout development as the "International Utility Tractor" or the "International All-Duty," but was renamed shortly before production. The tractor featured a lower profile for use in buildings and other low clearances (such as orchards), while being made more amenable to loaders. The tractor was also designed to be easier for the operator to get on and off; a feature useful on a "chore" tractor. The tractors found a ready market among both farm and industrial markets. Originally known as the "International 300" when released, the name was changed to "International 300 Utility" in mid-1955, with nameplates and serial number plates changed to match.

The Farmall 300 received as optional the Torque Amplifier and Independent PTO, as did the 300 Utility. The two features, first brought to production on the Super MTA and Super W-6TA,

A right-side view of the Farmall 300. The silver seat is the correct color for this vintage, and mufflers were made of aluminized steel. All steel belt pulleys were painted all red.

produced a small tractor that was very full-featured. The Farmall received Fast Hitch for the first time in the H-300 size tractor, and a more-advanced hydraulic system with integral lift system. The actual hitch points were larger than those on the Super C or 200, although the smaller-point implements could still be used with a special adapter. Fast Hitch was limited to tractors with two- or three-valve Hydra Touch hydraulic systems.

The Farmall and International 300 received the new C-169 engine. This engine had more power than the C-164 of the Super H, produced from an engine that had a $\frac{1}{16}$-inch-larger bore than the C-164's 3 ? inches. Revolutions per minute were raised from 1,650 to 1,750 in the Farmall 300 and some International 300s, while some other International 300s were rated at 2,000 rpm. A new carburetor was used to supply the extra fuel needed. The 300s were available with LPG, kerosene-distillate, and gasoline-fueled versions, and the Farmall 300 was available in the high clearance version whose components were little changed from those of the model HV high-clearance tractors that started production in the early 1940s.

Sheet metal was changed extensively from the Letter Series. The hood was of three sections, consisting of a left and right rear section and a front section. Instead of the hood line blending into the fuel tank like earlier tractors, the hood extended back into an instrument panel/reservoir housing, which also housed the hydraulic valves. The instrument panel housed the ignition and starting switch, instrument panel lights, battery charge indicator, oil pressure gauge, and heat indicator all in one location, rather than spread out all over the tractor as on the earlier H and Ms. Hydraulic levers were located on the right side of the panel on the Farmalls (to the right side of the operator on the Utilities), and the transmission-driven pump was eliminated as an option. Multiple hydraulic valves were used for single- or double-acting cylinders.

In early 1955, the International 300 Utility received power steering a new, special attachment that would rapidly become a favorite, especially for loader tractor. A Ross integral-type steering unit was used. Only tractors fitted with Hydra Touch (either one, two, or three valves) could be fitted with the new attachment, as hydraulics were needed to power the Ross unit. On the Utility, factory or field installation was possible.

Power steering attachments for the Farmall 300 and 400 Series tractors became available in early

1956. Behlen hydraulic power boosters were assembled as an integral part of the steering shaft. The system could only be used on tractors with Hydra Touch hydraulic systems. Other parts necessary were new regulators, flow valves, tubes, and attaching parts. The attachments were factory only when released.

The changes in the new 350 line, introduced in late 1956, foreshadowed many events to come. The chassis of the 350s, both Farmalls and Utilities, saw the new weight transfer Fast Hitch and new push-button starting (replacing the turnkey of the 300s), as well as the two-tone paint, hood sheet decal, and new serial number plates. The gas tractors received the new C-175 increased horsepower engine. The engine bore was increased 1/16 of an inch over the C-169 used in the 300s by using new, thin-wall cylinder sleeves and larger pistons and rings.

The diesel-powered tractors really covered new ground. IH went outside the company to purchase direct-starting diesels from Continental Motors. Though IH had produced small, direct-starting diesels in Germany since 1950, and had recently put similar engines into production in England, domestic IH engines were not ready for production. The decision authorizing the new Continentals calls the tractors "Direct Starting Diesel 350," leaving little doubt which new feature the Engineering and Sales Departments desired. The GD-193 engine was selected, which was a four-cylinder, 3 3/4 x 4 3/8 stroke operating at 1,750 rpm for the Farmalls and 2,000 rpm for the Utilities. The Continentals had a reputation of being hard to start, but soon new IH diesel engines would come into U.S. production.

Both diesel- and gasoline-burning 300 Utilities were offered from the start of production in an orchard version (both for factory and field application) using parts that were used in a field attachment for 300 Utilities.

All 350s and 350Ds were also built in Wheatland versions that had an increased ground clearance, heavy-duty front axle (similar to the Hi-Utilities), a new platform at the top of the main-frame cover similar to that of the International W-450, new fenders combining the crown of the W-450 with new side sheets, and new "Wheatland' decals.

The 350 Utilities also saw some new applications in specialty crops. Both diesel and carbureted versions were produced with a "Hi-Clear" attachment for shade tobacco producers.

Engine
Gas
The 300 Series tractors used C-169 carbureted engines that differed slightly between the Farmall 300, Farmall 300 HC, and International 300 tractors. Bore was 3 9/16, while stroke was 4 1/4 inches.

The 350 tractors used the C-175 engine that was 3 5/8 inch bore and 4 1/4 inch stroke. The engines were also available with kerosene, distillate, or LPG attachments.

The 5,000- and 8,000-foot high-altitude attachments were eliminated in November 1956, in favor of a single high-altitude attachment, which equated to the old 5,000-foot attachment.

Kerosene/Distillate
Kerosene and distillate attachments were available for all 300 and 350 tractors that were not diesels. Fuel adjustment decals 1000678R1 (English, 1000769R2 Spanish, 1000821R2 French) were used, as were gasoline decals 1000680R2 (English, 1000770R2 Spanish, 1000822R2 French) for the starting tank. Radiator shutters were used, as were hoods with holes to fit the starting gas tank.

In October 1955, changes were made to the kerosene/distillate attachments that involved the use of new lightweight, gray iron pistons and related parts, matching similar parts used in regular production. In January 1956, the 8,000-foot altitude attachments for kerosene/distillate engines were canceled due to low demand.

LPG
LPG-burning attachments were available for all 300 and 350 Series carbureted tractors (although LPG burning for 300 Utilities was not available until at least March 1956). The changes to the tractor were major—a large fuel tank, the hood cut out to surround it, and a different cylinder head and pistons to provide the higher compression necessary. Ensign model XG 1 1/4-inch updraft carburetors were used. To crank the engine over against the higher compression ratios, a 12-volt electrical system was installed. A different air cleaner was also installed.

Tractors with the attachment got two "LP Gas" nameplates. A decal for butane propane equipment (from Underwriters Laboratory), 1001500R1, was put on, as well as a "Caution-12 Volt Electrical System" decal, 1001722R1, an oil filter instruction decal (added July 23, 1957) and a warning decal for the 12-volt electrical system, 1000813R1. The 350 LPG Series tractors got a different hood sheet nameplate background decal

Table 4.1
Kinds And Codes
Kind 511 Farmall 300

Attachment	Code1 Dom	Code 2 Dom	Code 3 Dom	Code 4 Dom	Code 5 Dom	Code 6 Dom	Code 7 Dom	Code 8 Dom	Code 1 Exprt	Code 2 Exprt	Code 3 Exprt	Code 4 Exprt
354090R93 Pneumatic tire variable tread front wheel w/ 4.25 KA-16 DC Rim	x	x	x	x	x	x	x	x	x	x	x	x
354106R91 Front Wheel pneumatic tire & tube 5.50x16 4 ply F-2 tread	x	x	x	x	x	x	x	x	x	x	x	x
363024R93 Pneumatic tire rear wheel & W10-38 rim	x	x	x	x	x	x	x	x	x	x	x	x
354128R91 Rear wheel tire & tube wide base type 11-38 4 ply R-1 tread		x	x	x		x	x	x		x	x	x
354126R91 Rear Wheel tire & tube 10x38 4 ply R-1 tread	x				x				x			
357819R92 Deluxe type upholstered seat attachment		x	x	x		x	x	x		x	x	x
356890R93 Tilt back seat attachment		x	x	x		x	x	x		x	x	x
51071-DE Exhaust muffler	x	x	x	x	x	x	x	x	x	x	x	x
363065R92 Hydra-Touch (Double Valve)		x	x	x		x	x	x		x	x	x
363125R91 Cigarette Lighter	x	x	x	x	x	x	x	x	x	x	x	x
360637R91 Torque Amplifier (IPTO Type)		x	x				x	x			x	x
365268R91 Fixed Drawbar		x	x		x	x	x			x	x	x
363093R93 Kerosene burning										x	x	x
522857R91 Fast Hitch												x
362912R91 Fast Hitch hydraulic cylinder				x				x				x
363799R91 LPG Burning					x	x						
364715R91 LPG Burning							x	x				

363090R92 Distillate Burning attachment was optional with 363093R93 Kerosene attachment, eliminated March 5, 1955. Kind 4 export added same date. Domestic Codes 5,6,7,8 added January 3, 1956. 362955R92 Fast Hitch was replaced by 519306R92 February 10, 1956. Number changed only. 365268R91 Fixed Drawbar added 3-5-1956. 357819R92 Deluxe Upholstered seat and 356890R93 Tilt Back Seat bracket were combined into 366148R91 Deluxe seat and tilt back seat bracket 6-27-1956. 519306R92 Fast Hitch and 362912R91 Fast Hitch hyd. Cylinder were replaced by 522857R91 Fast Hitch and 366344R91 Fast Hitch hyd. Cylinder 7-18-1956

Table 4.2
Kind 514-Farmall 300 HC

Attachment	Code 1 Dom	Code 2 Dom	Code 3 Dom	Code 4 Dom	Code 5 Dom	Code 6 Dom	Code 1 Exprt	Code 2 Exprt	Code 3 Exprt
354091R91 Pneumatic tire variable front wheel w/4.50E-20 DC rim	x	x	x	x	x	x	x	x	x
356754R91 Front wheel tire and tube-6.00x20 6 ply I-1 Tire	x	x	x	x	x	x	x	x	x
363025R93 Pneumatic tire rear wheel and W10-38 rim	x	x	x	x	x	x	x	x	x
354130R91 Rear Wheel pneumatic tire and tube 11-38 6 ply R-2 tread	x	x	x	x	x	x	x	x	x
363820R91 LPG Burning				x	x				
364716R91 LPG Burning						x			
51071-DE Exhaust Muffler	x	x	x	x	x	x	x	x	x
360637R92 Torque Amplifier (IPTO Type)	x		x		x		x		x
365269R91 Fixed Drawbar	x	x	x	x	x	x	x	x	x
363065R93 Hydra-Touch (Double Valve)		x	x		x	x		x	x
363125R91 Cigarette Lighter	x	x	x	x	x	x	x	x	x
366148R91 Deluxe Seat &Tile Back Seat		x	x		x	x		x	x
363095R93 Kerosene Burning							x	x	x

359819R92 Deluxe type upholstered seat and 356890R93 Tilt back seat bracket were combined into 366148R91 6-27-1956. 360637R92 Torque Amplifier was eliminated from code 1 11-15-1954, Code 2 and 3 added at that time. 363125R91 Cigarette Lighter was added 12-2-1954. 359819R92 Deluxe type upholstered seat added 6-15-1955. Originally, 363095R93 Distillate burning was optional with the kerosene burning attachment for the export tractors, but was cut out March 5, 1955. 356890R93 tilt back seat added 9-9-1955. Code 4, 5, and 6 were added 1-3-1956. 365269R91 Fixed Drawbar added 3-5-1956

(2750025R1 lefthand, 2750026R1 righthand) to match the shorter hood.

Diesel

Diesels were added to the 350 tractors (both International and Farmall) in 1957. IH did not build the engines; Continental manufactured them as the GD-193 3¾ x 4⅜, four-cylinder engine. The engines were direct starting, avoiding the complicated carburetor and switch-over system IH had previously used. No starting crank was available for these tractors. A variety of changes were made

Table 4.4
Kind 816-Farmall 350

Code 1-Gas, Code 2 LPG, Code 3 Diesel, Code 4 350HC, Code 5 350 HC LPG, Code 6 Farmall 350 DHC
Code 1 Export Farmall 350, Code 3 Export Farmall 350D, Code 5 Export Farmall 350HC, Code 6 Export Farmall 350DHC

Attachment	Code1 Dom	Code 2 Dom	Code 3 Dom	Code 4 Dom	Code 5 Dom	Code 6 Dom	Code 1 Exprt	Code 3 Exprt	Code 4 Exprt	Code 6 Exprt
354090R93 Pneumatic tire variable tread front wheel 4.25 KA-16 DC Rim	x	x	x				x	x		
354106R91 Front wheel tire & tube-5.50x 16 4 ply F-2 tread	x	x	x				x	x		
363026R93 Pneumatic tire rear wheel & W11x38 rim	x	x	x				x	x		
354128R91 rear wheel tire & tube 12.4x38 4 ply R-1 tread	x	x	x				x	x		
365268R91 Fixed Drawbar	x	x	x				x	x		
51071-DE Exhaust muffler	x	x	x	x	x	x	x	x	x	x
363125R91 Cigarette lighter 6 volt	x			x			x		x	
363126R91 Cigarette lighter attachment 12 volt		x	x		x	x		x		x
363799R92 LPG burning		x			x					
365210R92 Kerosene Burning							x			
354091R91 Pneumatic tire variable tread front wheel with 4.50Ex20 DC Rim				x	x	x			x	x
356754R91 Front wheel pneumatic tire & tube 6.00x 20 6 ply I-1 Tread				x	x	x			x	x
363025R94 Pneumatic tire rear wheel & W10-38 rim				x	x	x			x	x
365269R91 Fixed Drawbar				x	x	x			x	x
365211R92 Kerosene Burning								x		
354130R91 Rear Wheel pneumatic tire & tube 11x38 6 ply R-2 tread				x	x	x			x	x

Table 4.3
Kind 548 International 300 Utility Tractor

Attachment	Code 1 Dom	Code 3 Dom	Code 1 Exp	Code 2 Exp
354106R91 Front Wheel pneumatic tire & Tube 5.50x 16 4 ply F-2 tread	x	x	x	x
361702R91 Pneumatic Tire Variable Tread Front wheel 4.25KA-16DC rim	x	x	x	x
361721R91 Rear Wheel Pneumatic tire & Tube 10x28 4 ply R-1 tread	x	x	x	x
361721R91 Pneumatic Tire rear wheel 10x28 rim	x	x	x	x
363125R91 Cigarette Lighter	x	x	x	x
363063R91 Hydra-Touch (Single Valve)	x	x	x	x
519401R92 Fast Hitch			x	x
366345R91 Fast Hitch hydraulic cylinder			x	x
364488R91 Rockshaft			x	x
363808R92 Tachometer	x	x	x	x
363832R91 Fixed Drawbar		x		

Domestic Code 2 became Code 3 April 29, 1955. Rockshaft attachment 362846R92 was replaced by 364488R91 May 5, 1955. Rock shaft eliminated, 519401R92 Fast Hitch was replaced by 522859R91, 363883R91 Fast Hitch Cylinder replaced by 366345R91 7-18-1956 (Weight Transfer type Fast Hitch) but was not put into production as a factory attachment for the 300.

in the regular 350 tractors to accommodate the new engine, including new air cleaners, front frame channels, redesigned controls, changed fuel tanks (to allow the use of return lines), 12-volt electrical systems, relocated outlet holes in the hood for air cleaners and exhausts, new radiator hoses, and new steering gear arms. On the 350D Utility, the toolbox was located on the top of the battery box, which was behind the seat.

Crank

Starting cranks were available for 300 and 350 carbureted tractors, and were mandatory for tractors shipped export. Any one of three cranks was used on the Farmalls (51544-DBX, 354705R31, or 360449R11), while the Internationals used 361924R11.

Oil Pan

Oil gauge drain cocks were used on kerosene/distillate tractors, while dipsticks were used in the other tractors.

Engine Exhaust System

The standard exhaust system for the Farmall 300s was exhaust pipe 361384R1, made of welded black steel pipe. This pipe was standard for the Farmall 350 carbureted tractors, but the diesels (including the International 350 diesel) used 366780R1, also steel pipe, which was a little longer and had some extra chamfering.

International 300 and 350 tractors had an under-mounted exhaust system with muffler as standard. Originally, muffler 362483R91 was used, but the change was authorized May 7, 1956,

Table 4.5
Kind 826-350 Utility

Kind 826 Code 14- 350D Utility with High-Utility Attachment
Kind 826 Code 15 International 350 Utility

Attachment	Code 3	Code 4	Code 5	Code 6	Code 12	Code 13	Code 14	Code 15	Code 16
368407R91 Hi-Utility tractor Attachment							x		
368406R91 International Hi-Utility tractor Attachment					x				
368408R91 International Hi-Utility tractor Attachment						x			
361702R91 Pneumatic Tire variable front wheel and 4.25-16DC rim attachment	x	x	x	x	x	x	x	x	x
354106R91 Front Wheel Pneumatic Tire and Inner Tube attachment 5.50x 16 4 ply F-2 Tread	x	x	x	x	x	x	x	x	x
363026R93 Pneumatic Tire Rear Wheel and w11-38 rim					x	x	x		
354128R91 Rear wheel pneumatic tire and inner tube attachment 12.4-38 4 ply r-1 tread					x	x	x		
365268R91 Fixed Drawbar attachment					x	x	x		
363126R91 Cigarette lighter attachment 12 Volt						x	x		
368903R92 Tachometer Attachment							x		
361720R91 Pneumatic tire rear wheel and 10-28 rim	x	x	x	x			x		x
363233R91 Rear Wheel pneumatic tire and inner tube attachment 11-28/ 12.4-28 4 ply R-1 Tread	x	x	x	x			x		x
363832R91 Fixed drawbar attachment	x	x	x	x				x	x
365126R91 Cigarette lighter attachment 12 volt	x		x	x			x		x
367142R91 Tachometer attachment	x			x					
363125R91 cigarette lighter attachment 6 volt		x			x		x		
363808R93 Tachometer attachment		x	x					x	x
364529R91 Front Power Take-off pulley attachment								x	x
365780R93 Liquid Petroleum gas burning attachment			x			x			x
367845R92 Wheatland Special Tractor		x	x						
368005R92 Wheatland Special Tractor Attachment				x					
363832R91 Fixed drawbar attachment		x	x						
365395R91 Vertical exhaust muffler attachment		x							
362834R92 Torque Amplifier attachment		x	x	x					
368902R92 Tachometer Attachment					x	x			

Front power take-off pulley attachment 364529R91 was added to the Kind 826 code 15 and 16 tractors 2-18-1958. Code 15 was originally Code 1, Code 16 was originally code 2, changed 4-14-1958, because of added front power take-off addition.

to use 365966R91 (produced by Donaldson or Mac Kenzie), which was part of a clutch improvement package but also reduced noise and heat. At the same time, muffler shield 365967R11 was added. These changes were effective on engine C-169 serial number 62483 and up.

Electrical
Standard were 6-volt electrical systems on carbureted tractors, with LPG tractors and 350 diesels getting 12-volt systems. Some International 350 Wheatland Specials also got 12-volt electrical systems. Starting and lighting systems were available as field attachments as well as factory attachments.

Turnkey electrical systems were used on the 300 series tractors, but a variety of problems forced the change to a key-lock, push-button starting system on the 350s. This required changes to wiring harnesses, instrument and control panels, and a variety of other parts.

Table 4.6
Piston, Cylinder Head, and Compression Ratio Chart, C-169 (300 Series Tractors)

	Piston	Cylinder Head	Compression Ratio	Manifold
Gasolene	364917R1	361479R1	6.8:1	362536R1 Farmall 361564R1 Utility
Gasolene, 5000 Foot Altitude	364996R1	361479R1	7.81:1	362536R1 Farmall 361564R1 Utility
Gasolene, 8000 Foot Altitude	365202R1	361479R1	8.56:1	362536R1 Farmall 361564R1 Utility
Distillate	365212R1	363257R1	4.75:1	358644R1
Distillate 5000 Foot Altitude	364999R1	363257R1	5.45:1	358644R1
Kerosene	365212R1	363258R1	4.50:1	358644R1
Kerosene 5000 Foot Altitude	364999R1	363258R1	5.10:1	358644R1
LP Gas	362731R1	364017R2	8:75:1	363788R1 Farmall, 36576R1 Utility

Made available for all 6-volt standard tractors in June 1957 were 12-volt electrical system attachments that could be applied in the field to older 350 Series tractors; 12-volt caution and warning decals were part of the packages.

The engine and accessories are all painted red, just like IH did it at the factory. IH produced their own carburetors at this time.

Battery
Auto Lite or Globe Union batteries were used on 300 and 350 Series tractors in production.

Magneto
IH's trusty H-4 magneto was available as an option on all the carbureted versions of the Farmall and International 300 and 350. Two specialized attachments for the Hi-Clear were tractors produced without starting or lighting equipment: one for domestic use and one for export use only (tractors shipped overseas had a starting crank as standard, while the domestic tractors had the crank as an attachment). The 300 Utilities with a magneto used a push-button starting switch rather than the key system used by the tractors with battery ignitions.

Headlights
Sealed beams were available in both 6-volt and 12-volt systems. The lights were the flat back type. Sealed beams were apparently illegal in Switzerland until 1955, so the old-fashioned teardrop headlights were used on Farmall 300s shipped there.

Combination Rear Light and Taillight
The 6-volt combination rear light and taillight (large clear lens, with a small red lens to one side) was available from the start of production. For those tractors that needed them, 12-volt lights became available and were subsequently installed on LPG and diesels when these tractors were produced.

In August 1956, a change was authorized to relocate the rear lamp to the lefthand fender instead of the righthand fender, so the lamp could be seen on the left side of the tractor at night when on the road and to clear the new Traction Control Fast Hitches. There was a field attachment so the lamp could be changed on older Farmall and International 300 tractors as well. The change involved a new clip and wiring harness.

Safety Light
Safety-light packages were offered as field attachments on these tractors as of March 21, 1955. Original lamps were 6-volt guide lamps number 897104 (IH package number 363914R91). On April 13, 1956, 12-volt guide lamps 897105 were added for tractors with 12-volt electrical systems, mainly diesel and LPG tractors. The packages were intended for tractors with fenders, but if fenders were not fitted, a safety lamp support was used. The tractors using the safety lamp, which could be taken off the tractor and put on the rear of an implement, had to be fitted with breakaway connector socket package 363923R91. The electrical system caution decal had to be fitted (1001722R1 12-volt or 1001725R1 6-volt). The lamps were two sided, with amber glass in front, and red in the rear.

Generators and Starters
Delco-Remy made the generators and starters.

Although many Farmall 300s had the two-point Fast Hitch, it was by no means mandatory, and many had the swinging drawbar like this one does. The live PTO handle is between the fender and seat running down to the clutch unit. Many people believe that you can add a live PTO by just adding this unit, but in reality the live PTO tractors have an almost entirely different transmission from the 300s without it.

Cigarette Lighter

Several different cigarette lighter attachments were available, depending on the tractor. Six-volt units were used on Farmall and International 300 tractors either factory or field, and were also used on gasoline versions of the 350. The 350 Series tractors with 12-volt electrical systems, LPG attachments, or diesel engines received 363126R91, a 12-volt lighter was authorized in October of 1956.

Chassis
Serial Number Plates

Serial number plate 0362395R1 was used on Farmall 300 and 300 HC. The plates were changed to R2 in March 1955, when the maximum idle speed and overload warning were taken off the plates. All 300 Utilities used 0363929R1. Farmall 350 series used serial number plate 0366682R1; the 350 utilities used 0366681R1.

Hood and Radiator Grille

Hoods varied by tractor, due to the different drilling necessary to accommodate the nameplates and for holes for auxiliary-starting gas tanks for kerosene and distillate tractors. Tractors shipped for export with different nameplates also had hoods specifically drilled.

Nameplates and variant plates were made of stainless steel with the exception of the "IH" symbol on the radiator, for which either stainless steel (362397R1) or die cast was used (365371R1,

added December 10, 1955). "McCormick" plates were used on some 300 Utilities shipped for export, while International nameplates were used on some Farmalls exported.

Radiator and Connections

The Utility radiator was several inches shorter than the Farmall radiator, among other differences. The overflow pipe was soldered to the top of the radiator in two places instead of three starting in June 1955 on 361703R92 (replacing R91). The goal was to reduce breakage of the pipe at the filler neck.

A close-up of the Torque Amplifier handle (all bare metal and red) and the Torque Amplifier nameplate in the background.

Originally, Stant Mfg. Co (361705R91) and A.C. Spark Plug radiator caps were used. On January 23, 1957, FTC 15506 changed the caps to one number (361705R91) and added Eaton Mfg. Co. as a supplier.

A radiator screen was used, having 26-gauge galvanized wire in a 5 x 5 mesh.

Fuel Tank Supports and Piping

A heat shield was added to protect the fuel tank, authorized in December 1956.

An auxiliary fuel tank, 366544R91, was added as a field parts accessory on International 300 and 350 gasoline-burning tractors in September 1956. The tank was mounted on the rear frame cover to the rear of the seat. The fuel tank support was integral to the tank, and was also used for the toolbox (which also mounted at that spot). The tank held about five and a half gallons; the fuel tank plumbed into the rest of the system at the sediment bulb.

Instrument Panels

Both the Farmall and International tractors had instrument panels, but the panels were quite different. The Farmall panels were one flat piece. The Internationals had the instruments on one panel, while a second panel (at an angle to the first) held the controls. On the 300 Series, the panels were painted with Tousey Varnish Company's 6278 Air Dry Varnish. The 350 Series had different instrument panels to accommodate the new push-button

Wheel rims of this era may have been galvanized or painted, depending on the manufacturer's fabrication practices and preferences. However, on the Farmalls, they were all basically silver/gray, with exact shade again depending on manufacturer. Wheel weights are a nice accessory, especially when the calcium chloride fluid weight is taken out of the tires to preserve the rims.

This tractor has power-adjust wheels, which were pioneered by Allis Chalmers and licensed to rim manufacturers. You loosen the clamps, then the wheels slide in and out as the tractor goes forward or backward, forced in or out by the angled bars the clamps rode on. The split weights are handier to move on or off the tractor—they weigh a bunch!

starting. The 350 series panels were painted IH-004-23 Satin Black.

Key ignition switches with starting were originally used in the 300 Series tractors; push-button starting was used on the 350. Two different push-button starting switches were authorized, 366316R91 or 366317R91, each having different manufacturers, apparently IH and Indak. The key remained, but did not activate the starter. Two ignition switches were used, carried under part number 366313R91: one by Indak Mfg. Co. and the other from Briggs and Stratton.

Instruments and Connections
The lighting switch knob, 0361997R1, was made of die cast zinc. The knob was painted with a light-gray enamel.

Air Cleaner Assembly
Donaldson air cleaners were used, with decals in the appropriate language of English, French, or Spanish. Air cleaners varied by fuel: gasoline and distillate-burning engines used cleaners with body 364492R91, LPG-burning tractors used 366452R91, and diesel-burning units used 306719R91.

Air cleaner hose clamps 45813-D were purchased from Wittek (their part number GM-53) and were used on Farmall 350D, 350 DHC, and International 350 D until a change was authorized November 23, 1956, when they were replaced by 39226-H (Wittek GM-56) due to a change in size. These were then replaced by 39233-H (Wittek GM-60), authorized January 18, 1957. 45813-D was also used on Farmall 300, 350, 300 HC, and 350 HC, and International 300 and 350 tractors as the air cleaner pipe to carburetor hose clamp starting sometime before March 30, 1955, replacing 39212-D.

Part number 45715-D (Wittek GM-45) was used as an air cleaner pipe to air cleaner hose clamp for all 300 and 350 series tractors starting March 30, 1955, when it replaced 39191-H (Wittek GM-40).

Clutches
IH and Rockford 10½-inch diameter clutches were used.

On the 300 Utility, changes to the clutch pedal took place on serial number 31927 that included a new clutch pedal, return spring, operating rod, and other parts to help reduce the effort needed to release the clutch. These changes required a change to the left platform (a cutout to provide

clearance for the new equipment), and new mufflers, exhaust pipes, and clamps, which again provided clearance, but also reduced noise and heat. A field package was available to put the changes into the older tractors.

This Farmall 300 had a three-valve Hydra-Touch system. The flat-back lights were used on the Hundred series tractors.

Torque Amplifier and Controls
Torque Amplifiers were available as special attachments for all tractors, with either provision for Independent power takeoff or transmission-driven power takeoff. There were variations in the attachments for the various tractors, International Utilities with different attachments from the Farmalls, and Farmall diesels different from the spark ignition versions. Most variations were due to control handle and linkage length differences (with changes also occurring between the 300 and 350 series as well). Changes were made in production to reduce the main clutch release effort, with field packages available for many of the changes. Trac-

Table 4.7
Variable Tread Front Wheel & Rims

Attachment number	Rim Size	Rim Manufacturer	Hub	Tractor Used With
354090R91	4.25KAx16DC	Goodyear Firestone Cleveland Welding French & Hecht Electric Wheel	8274DBX	F-300, F-350, F-350D
361702R91 (Disc Type)	4.25KAx16DC	Motor Wheel Kelsey Hayes Electric Wheel	Disc	I-300, I-350, I-350D
364492R91	5.50x x -16DC	Motor Wheel Electric Wheel	NA	I-350, I350-D

364492R91 Added March 19, 1955.

Table 4.8
Front Tire Attachments

Attachment number	Tire Size	Plies	Tread	Used with Rims	Tractor Used With
354106R91	5.50x16	4	F-2	4.25KAx16DC	F-300, I-300, F-350, F-350D, I-350, I-350D
354108R91	6.00x16	4	F-2	4.25KAx16DC	F-300, I-300, F-350, F-350D, I-350, I-350D
354109R91	6.00x16	6	F-2	4.25KAx16DC	F-300, I-300 F-350, F-350D, I-350, I-350D
356754R91	6.00x20	6	I-1	4.50Ex20DC	F-300HC, F-350HC, F-350DHC
355748R91	7.50x10	6	F-2	Single Front	F-300, F-350, F-350D
355436R91	7.50x16	8	I-1	Single Front 5.50F x 16DC	F-300, F-350, F-350D
364344R91	7.50x16	6	F-2	5.50F x16DC	I-300, I-350, I-350D
364656R91	7.50x16	6	I-1	5.50Fx16DC	I-300, I-350, I-350D
371412R91	7.50x16	10	I-1	5.50Fx16DC	I-350, I-350DC

364656R91 added July 1, 1955. 354108R91 had I-300 added April 19, 1955. 371412R91 added April 7, 1958.

Table 4.9
Variable Tread Front Wheel

Attachment number	Rim Size	Rim Manufacturer	Wheel	Tractor Used With
354091R92	4.50E-20DC	Goodyear Firestone Electric Wheel Cleveland Welding	9504-D	F-300HC, F-350HC, F-350DHC

tors fitted with Torque Amplifiers also received the "Torque Amplifier" nameplate (362394R1), which was attached with tubular clips.

Transmission

The transmission was the five speed based off the Farmall H. Tractors with the Independent PTO option had the Independent PTO shaft running inside the transmission main shaft. Two different bull-gear hubs were authorized for production: 0360539R1 (steel) or 0360540R1 (gray iron).

All tractors used the gearshift knob 53717-DA, which was made from black material but painted red in production.

In May 1955, changes were made in the gearshift lever handle end, part number 361670R1, that increased clearance between the steering wheel and gearshift ball. The part number suffix was advanced to R2.

Front Wheels

All wheel equipment was considered an attachment; there was no "standard" wheel or tire. If a dealer forgot to specify the equipment on an order, a default set of equipment was shipped. On Farmall 300, 350, and 350 D tractors the default front was the 354090R93 attachment with a 4.25 KA-16 DC rim. International 300, 350, and 350 D Utilities used the 361702R91 attachment with 4.25 KA-16 rims. The above tractors used as a default the 354106R91 front pneumatic attachment with 5.50x16 4-ply F-2 tread tires.

Farmall 300 HC, 350 HC, and 350 DHC default front wheels used the 354091R91 attachment with 4.50 e-20 DC rims, and the 356754R91 tire and tube attachment with 6.00x20 6-ply I-1 tread tires.

Steel wheels were available for the Farmall 300 and 350 tractors for factory or field use under attachment number 354134R91

Skid rings were available that were 2 ? inch high. These were purchased, and were carried under parts number 67893-DA or 53014-DA.

Either gray iron or sheet steel hubcaps were used until April 29, 1955 after which only sheet steel hubcaps were used. This was an elimination of an emergency substitute material part after the end of the Korean War.

Steering Gear

Two different steering wheels were available initially: 60070-D (made by Sheller Mfg. Co.) and 29118-DC (French & Hecht). Both 18 inches in diameter. These were replaced July 25, 1956, with wheel 366557R1 (Sheller) taking its place. In October 1956, U.S. Rubber Co. was added as a steering wheel manufacturer. The new wheels were of the "automotive" type, made for use with serrated steering shafts, and flat nuts versus the previous acorn nuts. They had a hard rubber composition with black enamel finish. The change actually took place on the transition to the 350 series tractors.

All 350 tractors used an IH-monogrammed steering wheel cap carried under part number 366558R1. This cap was actually manufactured by three different companies: Gits Molding Corp., Cruver Mfg. Co., and Greyer Molding Co.

Adjustable Front Axles

Wide front axles were available as both factory or field attachment on all Farmall 300, 350, and 350

D tractors. The diesels had a different fan pulley dust shield (part of the attachment).

The wide front axle attachment could not be used with M-448 or HM-639 cultivators, or HM-1 beet harvesters.

Apparently, there was a special front axle for Farmall 300, 350, and 350 D tractors for foreign-service repairs, 48540-DF. FTC-19602 authorized the use of this axle for domestic repairs in December 1958.

Single Front Wheel
Farmall 300, 350, and 350D tractors had available single-front wheel attachments. The attachments were available in two sizes, the 7.50x10 tire (6-ply) and the 7.50x16 tire (8-ply).

Brakes and Locks
Brake pedals differed between Farmall and Internationals due to the different seating positions. Pedals were made of malleable iron.

Rear Wheels and Tires
There were no standard rear wheels or tires; all were considered special attachments. If the dealer forgot to specify the rear wheel and tire equipment, the factory provided a default set. On the Farmall 300, the rear wheel and rim attachment was 363024R92 (W10-38 rims) and the tire attachment was 354128R91 (11x38 4-ply wide base R-1 tread tire and inner tube). The Farmall 350 and 350 D used 11x38 rims, and tire and tube attachment 354128R91, using 11x38 4-ply R-1 tread tires that were dual marked with the new industry standard marking of 12.4x38.

On the Farmall 300 HC, 350 HC, and 350 DHC, the default wheel attachment was 363025R92 or R93, which included the W10x38 rim. The rear tire and tube attachment was 354130R91 (11x38 6-ply R-2 tread tire).

On the International 300 Utility, 350 Utility, and 350 D Utility, the wheel attachment was 361720R91 (10x28 rim) and the tire and tube attachment was 361721R91 (10x28 4-ply R-1 tread tire).

Heavy-duty steel wheels were available with Farmall 300, 350, and 350 D tractors (not including Hi-Clears), either for factory or field application. The wheels were 51 inches in diameter with an 8-inch face, and were purchased from French & Hecht. When steel wheels were used, the two-keyway rear axle had to be used as of June 11, 1957. Extension tires were also available under attachment number 45236-DD.

Spade Lugs were available for use with the steel wheels, including 4-inch high by 3-inch wide, and 5-inch high by 3-inch high lugs. Two attachments (one of each size) were produced for use with extension tires.

Seat and Supports
Part 351880R92 was used as the base level seat for all 300 and 350 series tractors. It was a metal pan seat.

Trip rope clips were fitted to Farmalls, but not to International Utilities.

A tilt-back seat attachment was available on all Farmall 300 and 350 tractors, carried under attachment number 356890R93.

The International 300 Utility was IH's entry into the low-profile market pioneered by the Ford 9N. IH used the 300-series powertrain in a new, low chassis that featured easy on and off for the operator. These tractors were often used with loaders and were found on the farm and in town.

The rear lamp mounted was on the inside right fender on some 300s, and unlike the headlights, still used the classic teardrop shape. These wires would have been painted red in production.

Table 4.10
Rear Wheel & Rim

Some information after 1956 may be missing, also possibly some attachments.

Attachment number	Rim Size	Rim Manufacturer	Tire Size	Wheel	Used With
361720R92	W10x28	Goodyear Firestone Cleveland Welding French & Hecht Electric Wheel (Attached Clamp)	10x28 11x28	361724R1 (Motor Wheel, Electric Wheel)	I-300, I-350, I-350D
363024R92	W10x38	Goodyear Firestone Cleveland Welding French & Hecht Electric Wheel (Double Beaded)	10x38 11x38	363019R2	F-300, F-350, F-350D, I-350HC, I-350DHC
363025R92	W10x38	Goodyear Firestone Cleveland Welding French & Hecht Electric Wheel (Double Beaded)	11x38	363014R3	F-300HC, F-350HC, F-350DHC, F-350 (Heavy rear axle), F-350D (Heavy Rear Axle)
363026R92	W11x38	Goodyear Firestone Cleveland Welding French & Hecht Electric Wheel (Double Beaded)	12x38 11x38	363019R91	F-300, F-350, F-350D, I-350HC, I-350DHC
363027R92	W12x38	Goodyear Firestone Cleveland Welding French & Hecht Electric Wheel (Double Beaded)	13x38 12x38	363014R3	F-300HC, F-350HC, F-350DHC, F-350 (Heavy rear axle), F-350D (Heavy Rear Axle)
363028R92	W11x38	Goodyear Firestone Cleveland Welding French & Hecht Electric Wheel (Double Beaded)	12x38	363014R3	F-300HC, F-350HC, F-350DHC F-350 (Heavy rear axle), F-350D (Heavy Rear Axle)
363029R92	W12x38	Goodyear Firestone Cleveland Welding French & Hecht Electric Wheel (Double Beaded)	13x38 12x38	363019R2	F-300, F-350, F-350D, I350HC, I-350DHC
363987R91	W12x28	Goodyear Firestone Cleveland Welding French & Hecht Electric Wheel	12x28 13x28	361724R1 (Motor Wheel, Electric Wheel)	I-300, I-350D, I-350
364345R92	W12x24	Electric Wheel	14x24 13x24 12x24		I-300, I-350, I-350D
365493R92	W16x24	Electric Wheel	14x24		I-300, I-350, I-350D
369483R91	W12x24	Firestone Electric Wheel	14x24 13x24 12x24		I-350, I-350D
369484R91	W16x24	Electric Wheel	14x24		I-350, I-350D

364345R92, 365493R92 eliminated September 5, 1957, replaced with 369483R91, 369484R91

Deluxe-type seat attachments were available for all Farmall and International 300 Series tractors factory or field. Seats were made by either Milsco or Bostrom. Detachable seat pads with a molded underside to fit the seat were available as parts 351438R93 and 359483R91. All had silver Koroseal coverings during the 1954–1958 years, but later production used a white covering.

For International 300 and 350 Utility tractors with the deluxe cushion seat attachment, an arm pad attachment 361739R91 was available for factory or field installation. The pads were made by two different companies Milsco and Bostrum, under part number 363740R91 and had silver Koroseal coverings.

For a while, seats were used that had "Genuine Foam Rubber" tags sewn to them to stimulate sales. The foam rubber was replaced by polyurethane in May 1956.

A low seat attachment was available for 300 (after December 1955) and 350 Utilities. The seat mounted almost directly on the platform, and used seat 354157R92.

Hydraulics

It was possible to order tractors without hydraulic systems and add hydraulics later. IH did provide parts and assistance to add hydraulics systems to tractors in the field.

Hydra Touch System

Single-, double-, and triple-valve Hydra Touch systems were available factory or field. Handles were aluminum die-castings.

An auxiliary junction block was available for Farmall 300 and 350 tractors fitted with two- or three-valve Hydra Touch systems. The attachment was available as a field attachment starting in late December 1954.

Fast Hitch Hydraulic Equipment

Fast Hitch equipment was either a factory or field attachment for Farmall 300, 350, 350 D, and International 300, 350, and 350 D tractors. When the Fast Hitch was ordered from the factory, it was shipped with the tractor but not mounted on the tractor on the 300s. Double- or triple-valve Hydra Touch systems were required.

The Fast Hitch with weight transfer and articulated linkage was ready for production in July 1956, but it was delayed until the start of 350 production so the new features could be used as a sales point for the introduction. The weight transfer "Traction Control" could be retrofitted into 300 Series tractors.

Remote Control Equipment

Remote control adapter packages were authorized December 22, 1954, for tractors not fitted with them from the factory. A variety of packages was available, depending on whether the tractor had single-valve hydraulic systems, or two- or three-valve hydraulic systems, and whether the tractor had Fast Hitch or not. Farmall packages differed from International packages.

Hydraulic Pumps
Pesco and Thompson hydraulic pumps were used.

Attachments
Radiator Shutter and Controls
Radiator shutters were available for all 300 and 350 series tractors as a factory attachment. When fitted, a radiator shutter, control-rod operating-direction decal was applied (1000696R1 English, 1000763R1 Spanish, 1000815R1 French).

Fixed Drawbar
The fixed drawbar was an attachment on Farmall International 300 and 350 Utilities after November 29, 1955. Prior to this, they were standard. A drawbar extension plate was available as a field attachment.

Quick-Attachable Hi-Clear Drawbar
A quick-attach, high-hitch heavy-duty drawbar was available as an attachment for factory appli-

Table 4.11
Rear Tire Attachments

Attachment	Tire Size	Tread	Plies	Used with Rims	Tractor
354126R91	10x38	R1	4	W10-38	Farmall 300, 350, 350D, Int 350 and 350D W/High Clear
354128R91	11x38	R1	4	W11-38 (Note 1)	Farmall 300, 350, 350D, Int 350 and 350D W/High Clear
354129R91	11x38	R1	6	11-38	Farmall 300, 350, 350D, Int 350 and 350D W/High Clear
354130R91	11x38	R2	6	W10-38	Farmall 300, 300HC, 350, 350D, 350HC, 350DHC Int 350 and 350D W/High Clear
354789R91	12x38	R1	6	W12-38	Farmall 300, 300HC, 350, 350D, 350HC, 350DHC, Int 350 and 350D W/High Clear
354131R91	12x38	R2	6	W11-38	Farmall 300, 300HC
354790R91	13-38	R2	6	W12-38	Farmall 300 and 300HC
360585R91	12x38	R2	6	W11-38	Farmall 300
364348R91	14x24	R3	6	W12-24	International 300
364347R91	14x24	R1	6	W12-24	International 300
363998R91	12x28	R1	4	W12-28	International 300 Utility
363233R91	11x28	R1	4	W10-28, W12-28	International 300 Utility
364544R911	12x24	R1	6	W12-24	International 300 Utility
364356R91	11x28	R2	6	W10-28	International 300 Utility
361721R91	10x28	R1	4	W10-28	International 300 Utility
354550R91	13x24	R1	6	W12-24	International 300 Utility

Note 1 W10-38 was eliminated 5-23-1956, W11-38 added March 5, 1956, 354126R91 was eliminated April 7. 1958, International 350 and 350D added to tractors used March 4, 1957. 354129 had Farmall 350 HC and Farmall 350DHC added to tractors used June 6, 1957. 364356R91 added March 19, 1955. 354550R91 added June 1, 1955. 364544 added June 1, 1955. 364347R91 added March 19, 1955. 364348R91 added March 19, 1955. 363223R91 eliminated W12-28 rims June 8, 1955. 360585R91 added January 19, 1955.. 354126R91 eliminated April 7, 1958.

While loaders were a common use, International 300Us could also be fitted with two-point Fast Hitches and used with a variety of farm implements. The Utilities could also use the optional Independent Power Take Off, and they made dandy little tractors for dairy operations.

The International 300 had the same styling as the Farmall 300 and other IH tractors, but many sheet-metal parts were unique to the Utility. It featured good visibility forward.

cation or as a parts accessory for the Farmall 300 HC, 350 HC, and 350 DHC tractors throughout production.

Power Steering
Power steering for the Farmall 300 was announced in May 1956. The Internationals had power steering authorized in April 1955. The power steering could be either factory installed or applied in the field. The pressure for operating the system was supplied by the Hydra Touch pump or a front-mounted PTO pump on the International Utilities (as of May 1956). A flow-divider valve for the Hydra Touch pump provided power to the steering and implements, simultaneously.

Air Cleaner Extension Pipe
The air-pipe extension attachment (62970D) was available factory or field on the Farmall 300, 300 HC, 350, and 350 HC. It was also available as a parts accessory after the Farmall 350 left production. The attachment consisted of a long pipe with two braces.

Pre-Cleaner
Both the Donaldson (collector-type) and United Specialties (detachable sleeve–type) pre-cleaners were available on all Farmall 300 and 350 series tractors, factory or field, and also on the International 350 D Utility after November 5, 1956.

Pre-Screener

The Donaldson pre-screener was available on the Farmall 300 and 300 HC from the start of production. On November 20, 1956, the United Specialties pre-screener was authorized and both were available factory or field until the end of 350 production.

Exhaust Muffler, Pipe, and Spark Arrester

On Farmall 300 and 350 tractors, mufflers were an attachment carried under number 51071-DE. These mufflers were made of aluminized steel and had stamped into the shell the IH monogram and part number. An exhaust extension was carried under 53556-D.

Farmall 300s had available as a field or factory attachment an "orchard type" exhaust that was used most often as part of a cotton picker mounting attachment (all factory shipments were for cotton pickers). A cover was used in the hood hole, while the pipe and flange differed by whether the tractor was gas burning or kerosene/distillate burning. Both attachments used exhaust pipe 54567-D.

Farmall 300, 350, and 350 D had a spark arrester available, attachment 51579-D, throughout production. A different spark arrester was used for cotton picker tractors.

On International 300 utilities, the usual exhaust was the under-slung. In May 1955, the exhaust tail pipe was changed to incorporate a "whistle-type deflector" that directed the exhaust down and back, contacting the ground several inches from the inside of the rear lefthand tire.

International 300 and 350 tractors had available a vertical exhaust muffler as an option.

Belt Pulley

Belt pulley attachments were available for all 300 and 350 Series tractors. On the Utilities, however, the tractor had to be equipped either with a PTO or with a transmission equipped for PTO.

Several belt pulleys were available. The regular belt pulley was 9¾ inches in diameter with a 7½-inch face, either purchased from Browning Mfg. Co. or cast in gray iron by IH. Browning or IH pulleys were available in 11-inch diameter by 7½-inch face, 12-inch diameter by 89½-inch face. From Browning only, a 9¾-inch diameter with 8½-inch face, and a 6½-inch diameter by 8½-inch face pulley were available.

Cultivator Stay-Rod Anchor Brackets

Cultivator stay-rod anchor bracket attachments were available factory or field on Farmall 300, 350,

and 350 D tractors until September 17, 1956, when they became field parts accessories only.

Transmission-Driven PTO

Transmission-driven PTOs were available for all 300 Series tractors as a factory attachment, and as a field attachment for tractors that had the correct transmission option. When attached in the field or on a tractor with a Fast Hitch drawbar, drawbar extension plates had to be used. Plates varied by model.

The front axles were swept back to reduce the wheelbase to assist maneuverability. Axle damage can be an issue with tractors used with loaders—the axles will actually squat out when lifting a heavy load.

Independent Power Takeoff

An Independent power takeoff was available factory or field for all 300 and 350 Series tractors. Tractors that received the Independent PTO had to be fitted with the Independent PTO transmission, which could be ordered separately if the customer did not want to immediately pay for the Independent PTO when ordering a new tractor. Power from the engine went through the clutch

The instrument panel was made quite a bit differently from the Farmall. The power steering option was quite useful for a loader tractor and was identified in the steering wheel hub.

(but was not disconnected when the clutch was), then through the transmission main shaft and to a rear control unit that connected and disconnected power. Attachments varied for the particular tractor, also for tractors fitted with standard drawbars or Fast Hitch. When ordered for the 350 Wheatland Special, control rod assembly 367951R11 and control handle bracket 367960R1 had to be ordered.

Drawbar extension plates had to be ordered for tractors that were fitted with Independent PTOs in the field.

A service package on the operating lever rod pawl improved operation. When the PTO was in operation with a towed implement on a Fast Hitch drawbar, a Hydra Touch valve-lever lock was used.

Tachometer
Tachometers were available as a factory or field attachment for all 300 Series tractors, announced in May 1956. A variety were used due to different tire sizes and engine rpm.

Rockshaft
International 300 Utility tractors had available as a factory or a field attachment a rockshaft, carried under 362846R92. On May 5, 1955, this was changed to 364488R91, a rockshaft with increased hardness as well as a new two-piece depth-control bell crank.

Auxiliary Stay Rods
Auxiliary stay rod attachments were available for factory or field attachment for Farmall 300 HC, 350 HC, and 350 DHC tractors.

Tire Pumps
Engineair and Schrader tire pumps were available on all Farmall and International 300 family tractors, and continued on the 350s (except diesels).

Wheel Weights
Rear weights for the Farmall 300 and 350 tractors were part number 6818DA, weighing 145 pounds apiece. Up to three a side were used. International 300 and 350 Utilities used weight 362656R1, which also weighed 145 pounds, with up to two a side. International 300 Utilities with rear wheel attachment 365322R92 used a special split weight.

Front wheel weights were available for all 300 Series. Weights used on Farmall 300 and 350 tractors were 6788-D, weighing 42½-pounds each. Hi-Crops got weights 9042-D, weighing about 73 pounds each. Utilities 362655R1 weighed 54

Hydraulic systems were available with several valves. This tractor has the two-valve Hydra-Touch system. All were painted red.

The 450 was the largest IH row-crop tractor in the 1956–58 line, descended from the Farmall M. The grill was painted white, there was a white nameplate background decal on the side of the hood, and of course, the seats were silver.

The instrument panel for the 350. See the text for a description of the paint on the panel. Unfilled holes in the panel were originally covered by "buttons" that were a little bigger than the holes and had springs that held them in place. The 350s had a turn key and a starter button, while the 300s had a turn-key start.

Table 4.12

Extension Plate	Tractors Used
66426-D	F-300, F-350, F-350 D, I-300 with standard drawbar
66426-D	I-350, I-350 D with standard drawbar with or without Hi-Utility attachment
363445R1	F-300, F-350, F-350 D, I-300 with Fast Hitch drawbar
363445R1	I-350, I-350 D with Fast Hitch drawbar with or without Hi-Clear attachment
364013R2	I-300, I-350, I-350 D with Fast Hitch drawbar

pounds each. Each weight could be used up to two on a side.

Wheel weights were not for use with steel wheels.

Wide Tread Rear Axles

Longer rear axles were available as either a factory or field attachment (after May 1957 they became a parts accessory) to extend the rear tread. Originally, the maximum rear tread was listed at 100 inches but was changed to 105 inches (instead of the standard 93 inches), possibly before the start of 300 production.

The left side of the 350 gas, nicely detailed by Rick Wisnefske (and his remade dealer decal on the front grill). IH sprayed everything that was attached to the tractor at that point on the assembly line red, except for a few items that were masked off. There are more details in the paint chapter.

The two upper holes are for front-mounted cultivator braces. The larger tractors of the era enabled farmers to go to larger planters, which required wider cultivators that needed the braces. The seat cover is a later replacement.

Rear Axle Extensions

Farmall 300 and 350 Series had available rear axle extensions that attached with a cast steel clamp. The parts accessory or field attachment (it was not available as a factory attachment) was available for pneumatic wheel tractors only for use with single keyway rear axles. If a tractor had a double keyway axle, the single keyway axle had to be retrofitted.

Rear Wheel Fender (Farmall)

All Farmall 300 and 350 versions had available a rear fender attachment (51499-DB) that was available factory or field.

Front-Frame Channel Weights

Front-frame channel weights were available for tractors with Fast Hitch that weighed 125 pounds each, made in a left and right side.

Swinging Drawbar

Farmall 300s with Fast Hitch had attachment 363416R91. The attachment was available factory or field, but factory versions were not mounted at the factory. International Utility tractors with fixed drawbars (363832R91) had available factory or field swinging drawbar 364454R91. Farmall tractors without Fast Hitch received swinging drawbar 360631R91.

Farmall 350 and 350 D tractors with Fast Hitch and Traction Control received hitch 368451R91, while utilities with Traction Control received 368467R91.

Cotton Picker Mounts

Farmall 300 and 350 tractors were supplied for conversion to cotton picker tractors in a variety of variants. The modifications were either put on

Opposite: IH introduced diesel power for the first time in the smaller American Farmalls with the 350 diesel. It had an engine purchased from Continental that featured direct starting, instead of the start-on-gas system IH had previously used. IH was working on its own engines and had good direct-start engines for its production overseas, but they weren't quite ready for production on the 350.

This 350 has the weight-transfer Fast Hitch. If the tractor with a mounted tillage implement started to slip in the field, the operator could move the handle to transfer more of the implement's weight to the back wheels of the tractor to increase traction.

IH also provided kits to switch the tractors over in the field to normal operation. For cotton-picking season, the tractors operated in reverse, with the high drum tractors having the higher clearance parts installed. For the rest of the year, the tractor was operated front first (requiring a different shifter, rear cover, and flopping the drive bevel gear, and reversing the single front wheel), and high drum tractors were lowered. The high drum tractors had flanges on the rear axle carrier to which the drop housing for the high clearance feature bolted, which could be unbolted and replaced by a straight axle attachment. Regular Hi-Clears did not have these flanges located on the rear axle carrier.

Wheatland Special Tractor Attachment

As factory attachments, Wheatland Special packages could be applied to either I-350 or I-350 D Utility tractors. Fast Hitches were not used with the Wheatland packages. The tractors received a different front bolster and attaching hardware, an increased-clearance front axle, a different platform assembly with full coverage fenders, and "Wheatland Special" decals 2750074R1. Revised Independent PTOs, PTOs, and lighting attachments had to be designed to clear the new platforms and fenders; 24-inch rear tires could not be used with the package. Recommended were 13x28 rear tires with W12 rims and 6.00x16 front tires with 4.25KA rims. The first 350 Wheatlands, both diesel and carbureted, were built March 4, 1957.

Orchard Tractors

Orchard tractor attachments became available in mid-1956, both as field packages for 300 Utilities, and as factory attachments for 350 and 350 D tractors. There were two basic types with each package varying by whether Independent PTO was fitted or not, by tire size (in 12x28 and 14x24 tires with Independent PTO, and 12x28 tires only without), and by whether a cowl was used or not. Special platforms were used with these tractors.

Steering wheel shield packages were available separately for orchard tractors that did not already have them.

A new lower seat, designed specifically for orchard work was authorized in 1956. The seat was taken from the Farmall 100.

Hi-Utility

Hi-Utilities were 350 Utilities featuring Farmall 350 rear axles, wheels, and tire equipment, giving adjustable widths, and combined with modified

at IH's Memphis Works, or in the field as an attachment. The most familiar cotton picker tractors were the "high drum" variations, which included modifications for reverse operation and higher clearance (although these tractors are completely separate from "Hi-Clear" tractors that were produced for sugarcane and vegetable farming). The low drum tractors did not have the higher clearance modifications but did include reverse operation. The referenced drums are the cotton picker finger drums, the part that separated the cotton from the boll. Drum height was determined by whether the farmer was raising a shorter or a taller variety of cotton, which often varied by locality.

IH introduced Diesel power for the first time in the smaller American Farmalls with the 350 Diesel. It had a engine purchased from Continental that featured direct starting, instead of the start-on-gas system IH had previously used. IH was working on its own engines, and had good direct start engines for its production overseas, but they weren't quite ready for production on the 350.

front axle extensions and steering knuckles gave 5 inches of additional crop clearance. Different drive bevel gears yielded speeds the same as the regular Utilities, even with the larger rear tires. "Hi-Clear" nameplates were used in place of the standard "Utility" nameplates. Serial plates were modified by hand-stamping an "HC" and obliterating the "Utility." However, some confusion between the Utility Hi-Clears and the Farmall Hi-Clears resulted, so later in 1957 the names were changed to "Hi-Utility" on the nameplates, while the "HC" on the serial plate was changed to "HU." After the change, the "Utility" on the serial number plate was no longer obliterated. With the change in name, a restriction on sales to shade tobacco users was lifted.

Decals
1001182R1 decal, warning, drawbar and front pull hook (English, 1001184R1 Spanish, 1001185R1 French), (Farmalls only)
1001766 decal, warning, drawbar (English,

The right side of the Farmall 350 diesel. Unlike the larger IH diesel tractors, some of the 350 diesel's fuel filters were on the right side of the tractor.

Above: By the 1950s, competition in the tractor business was increasing. Diesel engines were becoming more accepted, and changes in technology and economics meant that smaller diesels were in demand, despite their higher purchase cost. While diesels were introduced primarily in the big small-grain fields of the West and Great Plains, corn farmers started to fall in love with them in the late 1940s. Right: Of course, IH made the large Farmalls in a High-Clear version as well. These tractors had a high-arched front axle and drop axles on the rear end (with chain drives). Due to their scarcity, these tractors are in high demand today. Randy Leffingwell

Far left: Front-wheel weights are a little less common than rear-wheel weights, but no less desirable. The large amount of torque a tractor puts out tends to lift the front end off the ground in heavy pulling conditions (or if the clutch is popped). Left: This Farmall also has power-adjust rear wheels that were easier to make small adjustments with compared to sliding the whole wheel in and out. Larger adjustments still had to be made at the axle. This tractor also has split-type rear-wheel weights.

1001767 Spanish, 1001768R1 French) 300 and 350 Series Utilities

2750019R1 decal, hood sheet nameplate background L.H. (Farmall 350 Series)

2750020R1 decal, hood sheet nameplate background R.H. (Farmall 350 Series)

2750015R1 decal, hood sheet nameplate background L.H. (International 350 Utility Series only)

2750016R1 decal, hood sheet nameplate background R.H. (International 350 Utility Series only)

1001735R1 decal, lighting switch (Farmall 350 Series?)

1000867R1 decal, lighting switch (International 350 Series?)

1000984R2 decal, warning, pressure cooling (English, 1000985R2 Spanish, 1000986R2 French)

1000636R5 decal, warning, power takeoff (English, 1000772R5 Spanish, 1000824R4 French)

1000704R4 decal, warning, brake (English, 1000767R4 Spanish, 1000819R3 French)

1001012R5 decal, instruction, oil filter (English, 1001020R5 Spanish, 1001102R5 French)

1001181R3 decal, instructions, air cleaner (English, 1001205 Spanish, 1001206R3 French), (Donaldson air cleaner)

1001289R1 decal, instruction, air cleaner (English, 1001290R1 Spanish, 1001291R1 French), (United Specialties Air Cleaner)

1000707R1 decal, gearshift

01001591R2 decal, patent marking

1001726R1 decal, caution, 6-volt electrical sys-

The gear shifters were curved to avoid hitting the seats and the operator. The knob was painted red in production. A nicely done instrument panel with about the right shade of paint.

tem (English, 1001726R1 Spanish, 1001727R1 French), (6-volt tractors only)

1001718R1 decal, "Made in the United States of America" (export only, including Canada, Mexico, Puerto Rico, and Cuba). Applied next to serial number plate.

1000813R1 decal, warning, 12-volt electrical system (English, 1000836R1 Spanish, 1000830R1 French)

2750122R1 decal, oil filter instruction (English,

The Farmall 350 High-Clear found a lot of uses. Originally designed for the sugarcane industry, High-Clears also found homes in vegetable crops and other special uses. Not many saw pastures. Randy Leffingwell

High-Clears often had to traverse terrain (sugarcane is grown in elevated beds), and that could mean trouble for a tall tractor. Optional auxiliary stay rods, running from the axle to behind the oil pan, help stabilize the tractor. Randy Leffingwell

7250123R1 Spanish, 2750124R1 French), (Farmall and International 350 D Series only)

2750125R1 decal, primary fuel filter instruction (English, 2750126R1 Spanish, 2750127R1 French), (Farmall and International 350 D Series only)

2750128R1 decal, secondary fuel filter instruction (English, 2750129R1 Spanish, 2750130R1 French), (Farmall and International 350D Series only)

2750131R1 decal, final stage fuel filter instruction (English, 2750132R1 Spanish, 2750133R1 French), (Farmall and International 350 D Series only)

Decal Changes
1001591R2 changed from R1 October 18, 1956, FTC-14991, addition or deletion of patent numbers.

Left: Cane tractors saw some heavy tillage use, and no doubt the Torque Amplifier came in handy. However, many cane tractors were part of fleet orders for use by hired help and were produced without a lot of attachments, so the TA is not anywhere close to universal on these rare tractors. Randy Leffingwell Below: Manifolds were painted red in production with the regular paint that burned off in use. Pretty much everything on the left side under the hood was painted. Hoods were painted separately and added later. Randy Leffingwell

Chapter Five
International 330

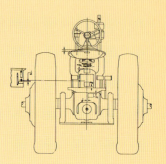

The International 330 Utility tractor is fairly rare; it was created when IH had an engine available well before the matching chassis for the International 340. IH put the new engine into the 350 Utility chassis to create a lower-price utility tractor until the new 1958 line was complete.

The International 330 was a tractor that had one foot in the old generation and one in the new. IH was planning a new generation of tractors for the late 1950s to replace the 130 through 650 families. Originally slated for 1956 or 1957, the program was delayed (probably because of financial problems) until 1958. However, tooling for the engine of what would become the 340-family gasoline engine was completed ahead of time, so IH decided to take the 350 Utility chassis and put the 340 engine in front of it. The resulting tractor was the 330. The tractor was produced for a short time, and was only moderately successful. The 340 engine was smaller than the 350, so the tractor had a reputation of being underpowered.

Engine
The engine was the C-135, having 135 ci in displacement, with a 3¼ inch bore and a 4¹⁄₁₆ inch stroke. Options available for the engine included an exhaust valve rotator attachment (factory or field), high altitude pistons (factory or field) and a front power takeoff attachment (factory only). The hydraulic pump drive was mounted on a separate pad below the ignition drive. All engines had dipsticks.

In production, the 330 had white wheels front and back, unlike most of the 350 Utility variations. The right side of the C-135 engine can be seen here.

The crankshaft was fitted with one of two nuts. One nut was intended for use with a front-mounted PTO, while another was for use with a starting crank.

Exhaust
Standard exhaust was under-slung. A short pipe was used at the end of the muffler to direct exhaust gasses down.

Carburetor
Zenith 68 x 7 carburetors were used in production. Marvel Schebler carburetors were authorized for production and service use with the C-135 engine in August 1959, after the 330 had finished production.

Air Cleaners
Donaldson and United Specialties were used.

Hydraulics
Hydraulics were considered attachments on these tractors. Some hydraulic parts were necessary to complete a tractor even if the hydraulics weren't ordered, specifically, the reservoir and some attached parts. Fast Hitch was available with the Hydra Touch system.

Chassis
As mentioned, the chassis was basically the 350 Utility chassis. See the 350 section for details.

Attachments
Attachments available included a foot accelerator,

Table 5.1
Kind 886- International 330

Attachment	Code 1 Export	Code 2 Domestic
361702R91 Variable tread front wheel w/ 4.25 KA-16 DC Rim	x	x
354106R91 Front Tire & Tube 5.50x16 4 ply F-2 tread	x	x
361720R91 Rear wheel w/ 10x28 tire	x	x
363233R91 Rear tire & tube 11x28 4 ply R-1 tread	x	x
363125R91 Cigarette Lighter (6 volt)	x	x
363832R91 Fixed Drawbar	x	x
369247R92 Tachometer	x	x
368664R91 Front Power Take-off Pulley	x	x

Table 5.2
Cylinder Head-Piston Commpression Ratio Chart

	Piston	Cylinder Head	Compression Ratio	Gasket	Manifold
Gasolene	367793R1	366206R1	7.37:1	366300R1	369645R1
High Altitude	368637R1	366206R1	8.48:1	366300R1	369645R1
Gas after 1960 (Service Parts)	374478R1	366206R1	7.60:1	366300R2	369645R1
High Altitude After 1960 (Service Parts)	374517R1	366206R1	8.70:1	366300R2	369645R1

The 330 had features of the 350 that had previously been unavailable in a smaller tractor, including the Independent Power Take Off and Torque Amplifier. The Fast Hitch had weight transfer like the 230 but was a more robust hitch.

The 330 was never available in a Farmall version, just the Utility version. The tractor was probably a little underpowered at the time given the large chassis, but commonality with the 350 ensured that a good range of equipment was available for this limited-production tractor.

The left side of the tractor shows the C-135 engine, the different air cleaner used on this tractor, and the underslung exhaust. Note the dealer decal under the headlight. The owner, Rick Wisnefske, had reproductions made of IH's 1950s dealer decal with his business name. IH made the decals available to its dealers with their individual imprint, but they were not mandatory. Note the correctly painted generator.

The right side of this tractor is detailed very closely to what IH probably did. The coils—as well as the distributor, etc.—on the few production photographs that have been found were painted very much like this.

arm pads, belt pulleys, front bolster weights, fuel gauge filler cap, cigarette lighter, fixed drawbar, swinging drawbar, electric lighting, Fast Hitch, Fast Hitch drawbar stabilizer, heavy-duty front axle, Hydra Touch, hydraulic power steering, front PTO hydraulic pump, Independent power takeoff, Independent PTO drive, H-4 magneto, vertical exhaust muffler, dual rear wheel mounting bolt and spacer packages, electrical breakaway connector, hydraulic remote control adapter, safety lamp, transmission-driven power takeoff, tire pump, pre-cleaner, pre-screener, reverse operation, seasonal PTO disconnect, low seat, deluxe type seat, tachometer, tires and tubes, Torque Amplifier, Torque Amplifier for Independent PTO, front and rear wheel weights, and wheels and rims.

Here's a 330 in original condition, about to go up for auction. These tractors are often found with loaders, which can be collectable in their own right. However, such tractors should be checked for front-end and steering damage.

Orchard Fenders

There were three different orchard fender configurations for International 330s. Parts accessory 366218R92 was a fender with cowling (sheet metal swept towards the engine) for use with 14x24 tires. Parts accessory 366221R92 was an enclosed round fender–only type without cowling for use with 14x24 tires. Parts accessory 366224R92 was a round enclosed fender without cowling for use with 12x28 tires.

Decals

1001181R3 decal, instructions, air cleaner (English, 1001205 Spanish, 1001206R3 French), (Donaldson air cleaner)

1001289R1 decal, instruction, air cleaner (English, 1001290R1 Spanish, 1001291R1 French), (United Specialties air cleaner).

1000704R4 decal, warning, brake (English, 1000767R4 Spanish, 1000819R3 French)

1001725 decal, caution, 6-volt electrical system

Table 5.3
Front Wheels & Rims Attachments

Attachment number	Rim Size	Rim Manufacturer	Wheel
361702R91	4.25KAx15DC	Motor Wheel Kelsey Hayes Electric Wheel	361714R91 (Disc Type)
364342R91	5.50Fx 16DC	Motor Wheel Electric Wheel French & Hecht	364343R91 (Disc Type)

Table 5.4
Front Tires and Tubes

Attachment number	Tire	Plies Size	Tread	Used with Rims
354106R91	5.50x16	4	F-2	4.25KAx16DC
354108R91	6.00x16	4	F-2	4.25KAx16DC
354109R91	6.00x16	6	F-2	4.25KAx16DC
364344R91	7.50x16	6	F-2	5.50Fx16DC
364656R91	7.50x16	6	I-1	5.50Fx16DC

Table 5.5
Rear Wheels and Rim Attachments

Attachment number	Rim Size	Rim Manufacturer	Tire Size	Wheel
361720R91	W10x28	Goodyear Firestone French & Hecht Electric Wheel	10x28 11x28	361724R1
363987R91	W12x28	Goodyear Firestone French & Hecht Electric Wheel	12x28 13x28	361724R1
369483R91	W12x24	Electric Wheel Firestone	12x24 13x24 14x24	369486R1*
369484R91	W16x24	Electric Wheel	14x24	369486R1*

Power adjusted rear wheels were used under part number 365323R92, made by Motor Wheel and Firestone.

Table 5.6
Rear Tires and Tubes

Attachment number	Tire Size	Plies	Tread	Used with Rims
361721R91	10x28	4	R-1	W10x28 W-Type w/attached clamp
363233R91	11x28	4	R-1	W10x28 attached clamp (Goodyear, Firestone, F&H)
363233R91	12x28	4	R-1	W12x28 attached clamps or attached rails
364632R91	11x28	4	R-3	W10x28 attached clamp (Goodyear, Firestone, F&H)
364347R91	14x24	6	R-1	W12x24 or W16x24 with disc attached
364348R91	14x24	6	R-3	W12x24 or W16x24 with disc attached
364356R91	11x28	6	R-2	W10x28 Goodyear with attached clamp
364547R91	12x24	6	R-1	Motor Wheel Rim & Disc
368370R91	12x28	6	R-3	with attached clamps
364550R91	13x24	6	R-1	Motor Wheel Rim & Disc
368007R91	13x28	6	R-1	with attached clamps
367657R91	12x28	6	R-1	with attached clamps or attached rails

(English, 1001726 Spanish, 1001727 French)

1001766R1 drawbar warning decal (English, 1001767R1 Spanish, 1001768R1 French)

1000707R1 decal, gearshift

2750015R1 decal, hood sheet nameplate background, L.H.

Torque Amplifiers are great for loader tractors, although they can be maintenance problems as well and should be checked for proper operation (or the repair bill calculated). This tractor has the original nameplate identifying it as having the optional Torque Amplifier.

2750016R1 decal, hood sheet nameplate background, R.H.

1000867R1 decal, lighting switch

1001718 decal, "Made in the United States of America" (export only, next to serial number plate)

1001012 decal, instructions, oil filter (English, 1001020R5 Spanish, 1001102 French)

1001591 patent marking

1000686R5 decal, warning power takeoff (English, 1000772R5 Spanish, 1000824 French)

1000984R2 decal, warning, pressure cooling (English, 1000985R2 Spanish, 1000986R2 French)

2750259R1 oil filler cap caution (English, 2750528R1 Spanish, 2750529R1 French), added to service package in January 1959

Electrall

All 330s could power two point and trail behind–type Electrall units, but to do so, they had to be fitted with a different PTO attachment with a 73-tooth drive gear and different shafts.

Chapter Six
Farmall 400, International 400, International W-400, Farmall 450, International W-450

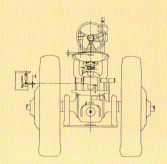

The Farmall 400 gas replaced the Super MTA in the top spot of the Farmall lineup. The tractor featured new styling, a new hydraulic system, and other improvements. It found good acceptance with row-crop customers.

The Farmall 400 had slightly more horsepower than the Super MTA it replaced. Compression ratios were raised slightly to 6.3:1 (gasoline version), and a new distributor assembly was used. Top engine rpm was dropped to 1,450 from 1,600, a popular selling point with salesmen who could point to lower wear and stresses. The Farmall 400 was available in kerosene/distillate-, gasoline-, LPG-, and diesel-burning versions, and was available in standard and high crop versions as well as serving as the power and chassis of high and low drum cotton pickers. The International 400 bore some resemblance to the International 300, but in fact it was close kin to the Super W6-TA. In May or June 1955, the International 400 became the International W-400. New nameplates and serial number plates were used, and a new front hood bracket and slightly different hood sheets were used to supply new holes for the plates. The

This tractor carried over the C-264 engine from its predecessor. The carburetor is painted red, as it was in production. The serial-number plate was moved to the other side of the clutch housing on the Farmall 400s, possibly because some mounted implements rubbed the plates on earlier tractors.

The data plate on the generator was masked off in production. The Torque Amplifier linkage is at the bottom—a popular, but not universal, feature on these tractors.

International 400 had power steering from the beginning of production, using a Monroe integral unit with an Eaton hydraulic pump working off the fan belt.

Sheet metal was changed extensively from previous Letter Series tractors. The styling and construction were very similar to the Farmall 300, maintaining the "line" look.

The Farmall 450 and W-450 saw a change to push-button starting; new C-281 and D-281 engines with increased horsepower; and the introduction of the new weight-transfer Fast Hitch,

Table 6.1
Kind 516 Farmall 400

Attachment	Dom Code1	Dom Code 2	Dom Code 3	Dom Code 4	Dom Code 5	Dom Code 6	Dom Code 7	Dom Code 8	Exp Code 1	Exp Code 2	Exp Code 3	Exp Code 4
354765R92 Variable tread front wheel w/ 4.25 KA-16DC Rim	x	x	x	x	x	x	x	x	x	x	x	x
354108R91 Front Tire and Tube 6.00x16 4 ply F-2 tread	x	x	x	x	x	x	x	x	x	x	x	x
363028R93 Rear wheel and W11-38 6 ply R-1 tread	x	x	x	x	x	x	x	x	x	x	x	x
354129R91 Rear Tire and Tube 11x38 6 ply R-1 tread	x				x				x			
354789R91 Rear Tire and Tube 12x38 6 ply R-1 Tread	x	x	x		x	x	x		x	x	x	
364582R93 LPG Burning					x	x	x	x				
357392R92 Exhaust Muffler	x	x	x	x	x	x	x		x	x	x	x
363450R92 Kerosene burning									x			
362874R92 Torque Amplifier (IPTO Type)		x	x			x	x				x	x
366148R91 Deluxe seat and tilt back seat bracket attachment		x	x	x		x	x			x	x	x
363053R93 Hydra-Touch System (Triple Valve)		x	x	x		x	x			x	x	x
363125R91 Cigarette Lighter	x	x	x	x	x	x	x	x	x	x	x	x
522858R91 Fast Hitch attachment				x			x					x
366244R91 Fast Hitch hydraulic cylinder				x			x					x
362940R92 Kerosene										x	x	x
365270R91 Fixed drawbar	x	x	x		x	x	x		x	x	x	

Code 4 added 12-2-1954. Tilt back seat bracket attachment 356890R93 added 12-2-1954. Codes 5,6,7,8 added 1-31-1955. LPG burning attachment 363470R91 replaced by 364582R92. Attachment 357810R92 Deluxe type seat attachment and 356890R93 Tilt back seat bracket attachment combined into 366148R91 6-27-1956. Fast Hitch attachment 362956R91 replaced by 519307R91 2-10-1956. Fast Hitch Cylinder 362212R91 was replaced by 363808R91 2-15-1956. 365270 Fixed Drawbar added 3-5-1956. Fast Hitch Attachment 519307R92, Fast Hitch Hydraulic Cylinder 362912R91 replaced by 5522858R91 Fast Hitch attachment, 366344R91 Fast Hitch Hydraulic Cylinder 7-19-1956.

Table 6.2
Kind 519 Farmall 400 HC

Attachment	Dom Code 1	Dom Code 2	Dom Code 3	Dom Code 4	Dom Code 5	Dom Code 6	Exp Code 1	Exp Code 2	Exp Code 3
354091R91 Variable tread front wheel w/ 4.50Ex20 DC Rim	x	x	x	x	x	x	x	x	x
356754R91 Front tire & Tube 6.00x20 6 ply I-1 tread	x	x	x	x	x	x	x	x	x
363028R93 rear wheel & W11-38 Rim	x			x			x		
363027R92 rear wheel & W12x38 rim		x	x		x	x		x	x
354131R91 rear tire & tube 12x38 6 ply R-2 tread	x			x			x		
354790R91 rear tire & tube 13-38 6 ply R-2 tread		x	x		x	x		x	x
357392R92 Exhaust Muffler	x	x	x	x	x	x	x	x	x
362874R92 Torque Amplifier (I.P.T.O.) type		x			x				x
363053R93 Hydra-Touch system (triple valve)		x	x		x	x		x	x
366148R91 Deluxe Seat & Tilt Back Seat Bracket		x	x		x	x		x	x
363125R91 Cigarette Lighter	x	x	x	x	x	x	x	x	x
363451R92 Kerosene Burning						x			
362942R92 Kerosene burning								x	x
364583R93 LPG Burning				x	x	x			
365269R91 Fixed Drawbar	x	x	x	x	x	x	x	x	x

Attachment 357810R92 Deluxe type seat attachment and 356890R93 Tilt back seat bracket attachment combined into 366148R91 6-27-1956. Domestic codes 5, 6, 7, 8 added 1-31-1955.

helpful in heavy traction situations. The new engines had a ⅛-inch increased bore over the C-264 engines used in the 400s. The extra bore was achieved by the use of new thin wall cylinder sleeves in the old block. The two-tone paint and new white hood sheet nameplate decal matched the rest of the IH line. The 450s retained the start on gas diesels and were, of course, also available with LPG fuel equipment. The Farmall 450s were also available in high-crop versions. The 450s were the real strength of IH's farm tractor sales as farms got larger and farmer numbers got smaller. How-

Table 6.3
Kind 521 Farmall 400 D

Attachment	Dom Code 1	Dom Code 2	Dom Code 3	Dom Code 4	Exp Code 1	Exp Code 2	Exp Code 3	Exp Code 4
354765R91 Front Wheel with 4.25KAx16 DC Rim	x	x	x	x	x	x	x	x
354108R91 Front Tire & Tube 6.00x16 4 ply F-2 tread	x	x	x	x	x	x	x	x
363028R93 rear wheel & W11x38 Rim	x	x	x		x	x	x	
354129R91 rear tire & Tube 11x38 6 ply R-1 tread	x				x			
354789R91 Rear Tire & Tube 12x38 6 ply R-2 tread		x	x	x		x	x	x
357392R92 Exhaust muffler	x	x	x	x	x	x	x	x
362874R92 Torque Amplifier (I.P.T.O. Type)			x	x			x	x
366148R91 Deluxe Seat and Tilt Back Seat Bracket		x	x	x		x	x	x
363057R92 Hydra-Touch System (Triple Valve)		x	x	x		x	x	x
363126R91 Cigarette lighter	x	x	x	x	x	x	x	x
522858R91 Fast Hitch				x				x
366344R91 Fast Hitch hydraulic cylinder				x				x
365270R91 Fixed Drawbar	x	x	x		x	x	x	

Attachment 357810R92 (Deluxe type seat) and 356890R93 (tilt back seat bracket) combined into 366148R91 6-27-1956. Fast Hitch Attachment 519307R92, Fast Hitch Hydraulic Cylinder 362912R91 replaced by 5522858R91 Fast Hitch attachment, 366344R91 Fast Hitch Hydraulic Cylinder 7-19-1956. 362956R92 Fast Hitch Cylinder replaced by 519307R922-10-1956. Code 4 Export added 3-5-1955. 365270R91 Fixed Drawbar added 3-5-1956.

Table 6.4
Kind 524 Farmall 400 DHC

Attachment	Dom Code 1	Dom Code 2	Dom Code 3	Exp Code 1	Exp Code 2	Exp Code 3
354091R91 variable tread front wheel w/ 4.50Wx20 DC Rim	x	x	x	x	x	x
356754R91 Front Tire & Tube 6.00x 20 6 ply I-1 tread	x	x	x	x	x	x
363028R93 rear wheel & W12x38 rim	x			x		
354131R91 rear tire & tube 12x38 6 ply R-2 tread	x			x		
363027R92 rear wheel & W12x38 rim		x	x		x	x
354790R01 rear tire & tube rear tire & tube 13x38 6 ply R-2 tread		x	x		x	x
357392R92 Exhaust muffler	x	x	x	x	x	x
366148R91 Deluxe seat & tilt back seat bracket		x	x		x	x
363057R93 Hydra Touch System (triple valve)		x	x		x	x
363126R91 Cigarette lighter	x	x	x	x	x	x
362874R92 Torque Amplifier (I.P.T.O. Type)			x			x
365269R91 Fixed Drawbar	x	x	x	x	x	x

Attachments 357810R92 (deluxe type seat) and 356890R93 (tilt back seat bracket) combined into 366148R91 6-27-1956.

Table 6.5
Kind 551 W-400

Attachment	Dom Code 7	Dom Code 8	Dom Code 9	Dom Code 10	Dom Code 11	Dom Code 12	Exp Code 7	Exp Code 8	Exp Code 9
364228R91 Front wheel 5.50F x 18DC rim	x	x	x	x	x	x	x	x	x
357984R91 front tire & tube 6.50x 18 4 ply F-2 tread	x	x	x	x	x	x	x	x	x
354930R91 rear wheel & DW12-30 rim	x	x	x	x	x	x	x	x	x
354941R91 tire & tube 13x30 6 ply wide base R-1 tread	x	x	x	x	x	x	x	x	x
357392R92 Exhaust Muffler	x	x	x	x	x	x	x	x	x
362875R92 Torque Amplifier (I.P.T.O. type)		x			x				
357819R92 Deluxe type upholstered seat		x	x		x	x		x	x
363039R92 Hydra-Touch (Single valve)		x	x		x	x		x	x
363125R91 Cigarette Lighter	x	x	x	x	x	x	x	x	x
364584R93 LPG Burning			x	x	x				
362943R92 Kerosene burning							x	x	x

Codes 4, 5, 6 were added 6-16-1955. 362875R92 Torque Amplifier, cut out of Code 4 and 5, 357819R92 Deluxe Seat, 363039R91 Hydra Touch removed from Code 4 7-22-1856. Codes 1, 2, 3, 4, 5, 6 were renamed Codes 7, 8, 9, 10, 11, 12.

Table 6.7
Kind 821 Farmall 450 & 450D

Domestic Code 7, 2 Export Code 9 Farmall 450, Domestic Code 5, 8 Export Code 10 Farmall 450 HC, Domestic Code 3 Farmall 450D, Domestic & Export Code 6 Farmall 450 DHC

Attachment	Dom Code 7	Dom Code 2	Dom Code 3	Dom Code 6	Dom Code 8	Dom Code 5	Exp Code 10	Exp Code 9	Exp Code 6
354765R92 Variable tread front wheel & 4.25 KA-16DC Rim	x	x	x					x	
354091R91 Variable tread front wheel w/ 4.50E-20DC rim				x	x	x	x		x
356754R91 Front Tire & Tube 6.00x20 6 ply I-1 tread				x	x	x	x		x
354108R91 Front tire & tube 6.00x16 4-ply F-2 tread	x	x	x					x	
363028R94 rear wheel & 11x38 rim	x	x	x	x	x	x	x	x	x
354129R91 rear tire & tube 12.4x38 6-ply R-1 tread	x		x		x			x	
354131R91 Rear Tire & Tube 12x38 6 ply R-2 tread		x		x		x	x		x
357392R92 Exhaust Muffler	x	x	x	x	x	x	x	x	x
363125R91 Cigarette Lighter (6 volt)	x				x		x	x	
365269R91 Fixed Drawbar			x	x	x	x			x
365270R91 Fixed Drawbar	x	x	x					x	
363126R91 Cigarette lighter (12 volt)		x	x	x		x			x
364583R95 LPG Burning		x				x			
365228R94 Kerosene burning							x	x	

Domestic Code 4 changed to Code 8 8-27-1957. 6 volt cigarette lighter replaced by 12 volt same date. Export Code 4 changed to Export Code 10 4-19-1958. Six volt cigarette lighter replaced by 12 volt 4-14-1958. Domestic Code 7 was Code 1 changed 8-27-1957. Six volt cigarette lighter replaced by 12 volt same date. Export Code 9 was Code 1, changed 4-19-1958, six volt lighter changed to 12 volt 4-14-1958.

ever, the increased horsepower started to result in an increased rate of rear-end failures, noticeable since at least the Super Series. This problem would worsen in the 560.

Although IH had used 12-volt electrical systems since 1941 for its diesel tractors, and since the early 1950s on its LPG tractors, in 1957 it finally got around to replacing the old 6-volt systems on some of the other carbureted tractors. The 450s received the welcome change—using many parts from the diesels and LPGs, but with a new cranking motor, ventilated generator, and cable harnesses. The battery ignition distributor and coil remained 6-volt, but with a resistor that cut out during starting to allow a full 12-volts to the coil.

Engine
Gasoline

The gasoline engines were the C-264 engine for the 400s, and the C-281 engine for the 450s, with 264 and 281 ci of displacement, respectively. Pistons were available in standard, and 5,000- and 8,000-foot-altitude compression ratios, respectively. Gasoline tractors had dipsticks for oil level measurement. The major differences between the C-264 and the C-281 were (on the C-281) thin wall cylinder sleeves, larger-diameter pistons and rings, a new carburetor, and a new battery ignition unit, as well as a revised manifold.

Table 6.6
Kind 553 W-400D (Domestic)

Attachment	Code 4 Exp	Code 5 Exp	Code 6 Exp	4 Dom	5 Dom	6 Dom
363228R91 Front Wheel w/ 5.50xF-18DC Rim	x	x	x	x	x	x
357984R91 Front Tire & Tube 6.50x 18 4-Ply F-2 Tread	x	x	x	x	x	x
354930R91 rear wheel & DW-12x30 rim	x	x	x	x	x	x
354941R91 rear tire & tube 14x30 6 ply wide base R-1 tread	x	x	x	x	x	x
357392R92 Exhaust muffler	x	x	x	x	x	x
363126R91 Cigarette lighter	x	x	x	x	x	x
357819R91 Deluxe Type Upholstered Seat	x	x		x	x	
363043R92 Hydra-Touch System (Single Valve)	x	x		x	x	
362875R92 Torque Amplifier (I.P.T.O. Type)			x			x

Codes 1, 2, 3 changed to 4, 5, 6 3-12-1956. New tractors had revised fenders.
IPTO taken out of codes 1,2 March 5, 1955.

Kerosene and Distillate Burning

Kerosene and distillate burning equipment was available for factory installation only. The engines were based on the gasoline C-264 engine for the 400s and the C-281 for the 450 Series. Heads and sleeve/cylinder assemblies provided lower compression ratios used with the fuels. Gray-iron exhaust-valve seat inserts were used. Exhaust manifolds had "heat boxes" that allowed exhaust gases

Table 6.8
Kind 831-W-450 & W-450D

Domestic Code 7 & 8, Export Code 7, W-450; Code 6 Domestic and Export- W-450D

Attachment	Dom Code 7	Dom Code 8	Exp Code 7	Dom Code 6	Exp Code 6
363228R91 Front wheel w/ 5.50F x18DC Rim	x	x	x	x	x
357984R91 Front Tire & Tube 6.5x 18 4 ply F-2 tread	x	x	x	x	x
367964R91 Rear wheel & DW 12x30 Rim	x	x	x	x	x
354941R91 Rear Tire & Tube 14x30 6 ply wide base R-1 tread	x	x	x	x	x
357392R92 Exhaust Muffler	x	x	x	x	x
364584R94 LPG Burning		x			
363126R91 Cigarette Lighter 12 volt	x	x	x	x	x
365233R94 Kerosene burning			x		

Code 4 was originally Code 1, changed 7-11-1957. Code 4 became Code 7 8-27-1957. Six volt cigarette lighter replaced with 12 volt cigarette lighter same date. Code 8 was originally Code 2, changed 7-11-1957. 354930R1 replaced with 367964R91 rear tire & DW12x30 rim 8-16-1957. Code 6 was Code 3, changed 8-28-1957.

The proper flat-back headlight is on this tractor. The Torque-Amplifier handle is painted red, just as in production. Tractors fitted with the TA had the nameplate put on, as seen just below the headlight.

Here's a view of the rear of the seat. It's critical to have the proper-size hardware attaching the seat to the battery box, otherwise the seat (and operator) can fall off. The couplings for rear-mounted hydraulics are located to the right of the battery box.

to heat the incoming fuel mixture to better vaporize the distillate and kerosene, which otherwise could result in poor economy and unburned fuel diluting the crankcase oil. Gray iron pistons were used. Pistons were available in standard, and 5,000- and 8,000-foot-altitude compression ratios.

In October 1955, changes were made to the kerosene/distillate attachments that involved the use of new lightweight gray iron pistons and related parts, matching similar parts used in regular production. In January 1956, the 8,000-foot-altitude attachments for kerosene/distillate engines were canceled due to low demand.

Tractors receiving the attachments got fuel adjustment decal 1000678R2 (English, 1000769R2 Spanish, 1000821R2 French) and the gasoline decal (1000680R2 English, 1000770R2 Spanish, 1000822 French). Hoods were different to accommodate the gasoline starting tank. Radiator shutters were mandatory; shutter assembly 363381R91 was used on tractors with two or more valve Hydra Touch systems, while 363379R91 was used on tractors with one valve or no hydraulic system. Fuel piping and the strainer assembly were different, since the strainer took fuel from two different tanks. The oil pan used "oil-gauge valve assemblies" (otherwise known as trycocks) that allowed the operator to drain unburned fuel that collected in the oil pan before it ruined lubrication.

LPG

LPG tractors used the C-264 and C-281 gas engine for bases during the production of each. Engine differences were many. The LPG tractors used a higher compression ratio than gasoline tractors, so sleeves, pistons, and heads were different.

Much of the rest of the tractor was different, mainly to accommodate the fuel tank. Originally the fuel tanks were all 24 inches in diameter, but on June 10, 1955, 22-inch outside diameter tanks were authorized. On June 22, 1955, the 24-inch tanks were eliminated. The different-sized tanks required different hoods to match the sizes. Tanks were manufactured by a variety of companies, including Western Engineering, Pressed Steel Tank, Eastern Tank, and Santa Fe Engineering. Carburetors were 1¼-inch-updraft Ensign Model XG.

Because of higher compression ratios, 12-volt electrical systems were used in all LPG tractors in the 400 series. Steering was a little different on the Farmalls, due to the need to run the shafting around the tank.

LPG tractors received certain decals and identification relating to the equipment. A

The Farmall 400 diesel carried on the Farmall diesel tradition with its start-on-gas engine that had a reputation for mechanical complexity but good cold-starting performance.

Into the 1960s many dairy farms in the Upper Midwest grew small grains and row crops for their cows' forage. The small pull-type combines are hard to find in good condition and are starting to get collectors' attention.

butane/propane equipment decal, 1001500R1, was adhered. The 12-volt electrical system decal was part of the attachment until May 20, 1957, when all 400 Series tractors went 12-volt, (1001720 vs. 1000815R1). Oil filter instruction decal 2750225R1 was added July 23, 1957. On the 450 Series tractors, the hood sheet nameplate background decals were different because of the different construction of the hood. LPG-burning tractors also received the "L.P. Gas" nameplate and a weather cap.

C-281 Trial Lot
A trial lot of 1,000 engines was put into F-450, F-450 HC, and W-450 tractors starting in late 1957. The engines were fitted with exhaust valve 0369624R1, which were cast, instead of having a Stellite facing. The valves had the manufacturer's trademark, the part number, the IH symbol, and "RMC" (Rich Mfg. Co.) etched on the head of each valve.

Manifold Heater Blocks
F-400, F-400 HC, and I-400 tractors with gasoline manifold 362437R1 or R2 could be fitted with a manifold heater block as a parts accessory. The block was part 367100R1, made of gray iron.

High Altitude Piston Attachments
A variety of high-altitude piston attachments was

83

This appears to be a fairly original instrument panel on one of the first thousand tractors built. The knob on the lower left is the cigarette lighter.

Oil gauges differed between diesel and gas tractors, and the ammeter worked on a 12-volt instead of a 6-volt system.

Table 6.9
Piston, Cylinder head, and Compression Ratio Chart, C-264 (400 Series Tractors)

	Piston	Cylinder Head	Compression Ratio	Manifold
Gasoline	357613R1 356965R1	362174R1	6.3:1	362437R1
Gasoline, 5,000-Foot Altitude	357238R1 357620R1	362174R1	7.2:1	362437R1
Gasoline, 8,000-Foot Altitude	357242R1 357623R1	362174R1	7.8:1	362437R1
Distillate	357226R1	362920R1	4.62:1	363153R1
Distillate 5,000-Foot Altitude	357316R1	362920R1	5.36:1	363153R1
Distillate 8,000-Foot Altitude	357319R1	362920R1	5.89:1	363153R1
Kerosene	357226R1	362921R1	4.3:1	363153R1
Kerosene 5,000-Foot Altitude	357316R1	362921R1	4.91:1	363153R1
Kerosene 8,000-Foot Altitude	357319R1	362921R1	5.33:1	363153R1
LP Gas	357262R2 357623R2	362174R1	7.8:1	358308R1
LP Gas*	357613R1 356965R1	364586R1	8.35:1	358308R1

*Added September 6, 1955, new cylinder head.

Table 6.10
Piston, Cylinder Head, and Compression Ratio Chart, C-281

	Piston	Cylinder Head	Compression Ratio	Gasket	Manifold
Gasoline	366483R1	362174R1	6.61:1	48957-DE	362437R2
Gasoline High Altitude	366521R1	362174R1	7.6:1	48957-DE	363437R1
Distillate	366497R1	362920R1	4.5:1	48957-DE	363153R1
Distillate High Altitude	366527R1	362920R1	5.17:1	48957-DE	363153R1
Kerosene	366504R1	362921R1	4.3:1	48957-DE	363153R1
Kerosene High Altitude	366533R1	362921R1	4.92:1	48957-DE	363153R1
LPG	366483R1	364586R1	8.5:1*	48957-DE	358308R1

*LPG compression ratio lowered from 8.76:1 to 8.5:1 March 27, 1957.

available for the carbureted tractors. Originally, attachments were available factory or field for 5,000-foot-altitude pistons and 8,000-foot-altitude pistons, but in late 1956, apparently the 8,000-foot-altitude pistons were eliminated and the 5,000-foot-altitude pistons were then referred to as "high altitude" pistons.

Diesel
The diesel versions used the same start on gas starting system used by IH since 1933. The 400 Series tractors used the D-264 engine (264 of course being cubic inch displacement) while the 450 series used the D-281. The diesels had a positive valve rotator attachment approved August 30, 1955. diesels were popular on export tractors, although domestic consumption was lower.

Air Cleaner
United Specialties or Donaldson air cleaners were used.

Starting Cranks
Starting cranks were shipped with tractors going overseas and were available as service items for domestic tractors. Diesel starting crankshafts (inside the upper bolster) were different from carbureted Farmalls, while the International carbureted and diesel-burning tractors also had separate crankshafts. The cranks also were different among the varieties of tractors, but in addition, several different cranks—apparently dating from various older tractors—were also authorized, presumably so dealers could clean out older stocks on their shelves.

The left side of this diesel engine has all the fuel filters and injection pump components you'd expect to see on a diesel, as well as numerous instruction decals. See the chapter on paint and decals for the precise location of all decals.

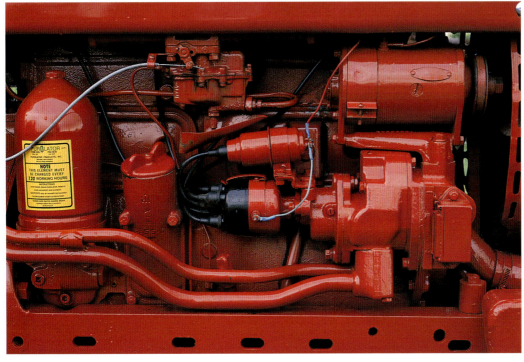

The right side of this IH-built start-on-gas D-264 has, unlike nearly every other diesel engine, a carburetor and an ignition. Note the inclined drip shield between the small carburetor on top and the ignition coil, which was designed to prevent a fire from gas dripping onto the high-voltage wiring. The two lines run to the hydraulic pump.

Exhaust

The base level exhaust was 361384R1 for the carbureted tractors (steel pipe) and 361385R1 for the diesels (gray iron pipe).

Electrical

Tractors shipped to Switzerland had the older style teardrop lamps until May 26, 1955. The flat back–style Guide Lamp lights were used on W-400 Series tractors from 501 to 3331, and on all Farmall tractors. On International W-400 serial 3332 and up and all 450s, a round back sealed beam headlight manufactured by Dietz was used as well.

The electrical system on all diesel and LPG tractors was 12-volt. On gasoline, kerosene, and distillate tractors, 6-volt was used until Farmall 450 serial number 11083, and W-450 1350.

This tractor had the optional frame channel weights, which aren't very common today. It would be a few years before the tractor industry would discover the now nearly universal front-mounted suitcase weight.

Electrical breakaway sockets were made by Cole Hersey Co. or Joseph Pollack Corp. The rear lamp and breakaway connector wiring harness was made by Packard Electric, Essex Wire, or Whitaker Cable, as were a variety of wiring harnesses used on the 400 and 450 Series tractors.

Generators were Delco-Remy; batteries were Globe Union or Auto-Lite.

Wiring harnesses were covered with impregnated woven braid.

The 400 Series tractors used a turnkey starting system. Because of certain problems with this system, the 450 Series tractors went to a key-lock with a push-button starting system. This required a variety of changes to the instrument panel, wiring harness, and other parts.

Cigarette Lighter

Cigarette lighter attachments were available for all tractors having electrical systems, although the point of attachment varied. Originally, the attachment was available factory or field, then just factory, then factory and field again, then the factory attachments were eliminated! The lighters were available in 6-volt systems initially, with a 12-volt system authorized in October 1956. Field attachments were available for older 12-volt tractors.

Chassis
Radiator

The radiator was the same for the W-400s, but the fan housing and the radiator inlet tube were different between the carbureted and the diesel-burning tractors. The inlet tube on the diesel was 2¾ inches long, versus 5⅛ on the carbureted tractors.

Radiator caps were carried under part number 361705R91, with three caps authorized for use: Stant Mfg. Co., A.C. Spark Plug Co., and Eaton Mfg. Co. The caps varied by handles and printing.

The material used for the radiator screen (on the inside of the grille) was galvanized steel #26 gauge wire in a 5 x 5 mesh.

Governor Controls

The sectors on the governor controls on the International W carbureted tractors were different from the diesel tractors. Part 363426R1 (carbureted) had 25 notches while 363743R1 (diesel) had 12 notches. The W-450 series tractors had a shorter governor control handle to clear different steering assembly used on those tractors. Farmalls used the same sector, diesel or gasoline.

Engine Controls and Instruments

The 400 and 450 Series tractors had an instrument and control panel. The top of the panel was rounded, with the gauges located over the controls. The choke was located to the left of the panel, supported by a sheet metal clip.

Sheet Metal

Hoods were made from two different blanks, one for gas, and one for diesel, for all 400 and 450 series tractors except LPG (which may have been made from the gas sheet). Beyond that, each version of the tractor had its own hood sheet, drilled differently for nameplates, exhausts, tanks, etc. The hardware used to attach the hoods were Cr-Rec-Truss head, 5/16 x 18 x ⅝ inch, with internal tooth lock washer screws.

The instrument panels were not painted red like the rest of the tractor. The 400 Series panels

With the Fast-Hitch system, the latches locked automatically when backing into the equipment, without the operator ever having to leave the seat. To unhitch the implement, this handle could be used from the seat to lift the latches on each side. It's nothing fancy, just a handle and a long rod that's hooked at the end.

Disc brakes were used throughout 400 and 450 production. They work great when maintained, but regular service is recommended.

were painted with Tousey's, 6278, air-dry enamel. The 450 Series panels were painted IH Satin Black.

Nameplates were made of stainless steel, except for some of the "IH" hood emblems, which could be either stainless or die cast after December 20, 1955 (all were originally stainless). On the W-series, the "W" plate was actually separate from the "450" due mainly to the W being added to the model designation later in production (the plates were authorized March 15, 1955).

The Farmalls used a different serial number plate than the Internationals, but all used #11 gauge by ⅜-inch-long, round-head brass escutcheon pins (there were even two different types of these). The serial number plates on all tractors were modified by a decision authorized March 7, 1955, which eliminated the maximum speed idle number and the overload warning, and changed the part number of the plates to R2.

The McCormick nameplates used on Farmalls (and on certain W-450s shipped export) were changed from part 362317R1 to 366685R1 August 30, 1956. The difference in the parts was that the new plate had a white background, where the old plate was red.

Fuel Tank

The standard flat fuel cap was 361910, although the author strongly recommends using the new Navistar-offered safety gas cap for tractors operating with gasoline.

The start-on-gas diesel tractors had a small auxiliary fuel tank for the gas, similar to those on the kerosene-distillate tractors.

The ram on the left provides the lifting power for the Fast Hitch, while the crank on the right helps adjust the hitch. The hydraulic couplers did have IH caps to cover them when not in use.

The loop on top of the Fast-Hitch receiver is the part of the latch that the previously mentioned hook grabs to release the implement. The Fast Hitch has a number of pins that are inserted or removed to configure the hitch for specific implements and conditions. If you're using the hitch, an operator's manual is highly recommended.

Two fuel tanks were used on the carbureted tractors. The original tank was 18 gallons capacity. Another tank was available with 21½ gallons capacity. If a gauge cap was purchased (see the attachment portion of this chapter) the appropriate gauge had to be used because tank depth was different, as was some of the piping.

Platform and Fenders
The International W-400 and W-450 Series tractors had platforms and full fenders as standard. The 400s had a low platform and narrow fenders until early 1956 (the change was authorized January 26, 1956). On the later tractors a higher platform and fuller, wider coverage fenders were used, including 400s and 450s. The original, fuller-coverage fenders were 365461R91 (lefthand) and 365462R91 (righthand) but these were changed to 367871R91 and 367710R91 on December 6, 1956. A conversion package was available from IH to convert the early W-400s and W-400 to the newer style fenders and platform. Whether factory or field, the electrical breakaway connector cables also changed to match the new fender construction.

Steering
The Farmalls used almost an entirely different steering system than the Internationals. Ross Mfg.

The large Farmalls were not shy about consuming fuel, and the regular gasoline-powered tractors were starting to decline in popularity due to fuel cost. LPG tractors were purchased in some numbers but were ultimately replaced by diesels due to a variety of reasons, including easier refueling.

LPG tractors usually comprised about 10 percent of production. One benefit of LPG was less lubricating oil contamination. To take advantage of the higher octane rating of LPG, these tractors had parts, including cylinder heads, that would run at higher compression ratios. Darius Harms is at the wheel.

Cylinder heads from LPG tractors are highly sought after by tractor pullers today for use in regular gasoline-powered M through 450 tractors. As a result, the small number of LPG tractors is getting smaller. The LPG-burning equipment required regulators and a carburetor, which were purchased from Ensign.

The tanks were not made by IH, but purchased by them from outside specialty manufacturers. They are a pressure vessel and should be inspected for safety before putting fuel in. Talk to your LPG suppliers.

Co. made the manual steering units on the Internationals. Steering gear assembly 363877R92 was replaced by 366563R91 July 25, 1956, effective for the change to the W-450 series.

The monogrammed steering wheel caps of the 450 Series were the same between tractors officially, although there was considerable variance in the actual caps. On the Farmall 400 Series the steering wheel was either part 60070D (Sheller) or 29118DC (French & Hecht) for the Farmalls (18-inch diameter, with horn). No cap was used. Sheller made the 363163R91 (20-inch diameter) for the International W-400 Series tractors. Bare steel parts of this wheel were painted red. On the all-450 Series, the standard steering wheel (less horn), 366557R1 was 18 inches in diameter, made by either Sheller or U.S. Rubber Co. On the 450 Series, the wheel and shaft were serrated, and a flat nut was used, versus the acorn nut of the 400 series. Part 366558R1 was the standard cap used for tractors with power steering, but there were three suppliers of that cap: Gits Molding Corp., Cruver Mfg. Co, and Breyer Molding Corp., which also made caps under part number 366566R1, used on 450 Series tractors without power steering.

Internationals fitted with power steering got a "Power Steering" nameplate, 366215R1.

Clutch

Clutches were different for Farmalls and Internationals. IH's own clutch was optional with a Rockford clutch. All were 12-inch diameter.

Transmissions and Final Drives

All tractors had the five-speed transmission that had evolved from the H & M transmissions designed in the late 1930s. However, horsepower and torque going through the transmissions had increased, so changes had been made in gear design and other areas. While most people think that the 460/560 tractors had reached some kind of "limit" in horsepower that resulted in failures, there is at least some evidence that 400s and 450s also experienced more than their share of transmission problems. One effect of the 460/560 problems was that service parts for the 400 Series tractors were often upgraded when the parts for the later tractors were upgraded as well, since either the original parts were the same or the parts were made using the same tooling.

The Super MTA and Super W6-TA introduced the Torque Amplifiers, which were developed into the Farmall 400. The TAs were an attachment, although they were usually included in the Kind and Code sales packages, so a large proportion of the IH 300, 350, 400, and 450 Series tractors carried them. TAs came in two kinds: one fitted for Independent power takeoffs (the shafts had to run through the area that the TA occupied) and one fitted for transmission-driven PTOs, which didn't have the extra shafting. Handles for the TA differed by whether the tractor was a Farmall or International standard-type tractor.

In the late 1950s and into the 1960s, available service packages had rear frames that included updates made in those parts that were also used in 560 and 660 production. These packages changed frequently when IH was struggling to come up with a fix on the later tractors.

When PTOs were not fitted, a gray iron cover was used (358841R1).

Farmalls and Internationals used different rear frame covers, with service parts being slightly different from production parts and having different parts numbers.

The third- and fourth-speed driving gears on the countershaft each had two different gears used in production.

Farmall 400, 400 D, 450, and 450 D tractors used transmission main shaft and drive bevel 3588611R1 or R2, while the Farmall 400 HC, 400 DHC, 450 HC, 450 DHC, and all the Internationals used shaft and bevel 358897R1 or R2. The shafts had slightly different length dimensions in several areas.

Regular Farmalls used a 50-tooth drive bevel gear in the differential, while Internationals and Hi-Clears shared a 47-tooth gear. Bull gear hubs could be either forged steel (although it seems none of these were actually made) or cast of gray iron until March 31, 1958, when the steel hubs were canceled.

Gearshift knob 53717-DA was used throughout production on all 400 Series tractors.

Brakes

Double disc brakes were used on all 400 and 450 Series tractors, with many parts purchased from Auto Specialties. The brake discs changed several times. Original discs were 357100R91; they changed to 366174R91 authorized on June 28, 1956. The disc part was changed to 368178R91 on January 28, 1957.

Front Frames

Service frame channels were the same for all 400 and 450 tractors, coming in left-hand- and right-hand varieties. However, in production, the diesel and carbureted versions were slightly different, made from the same part but drilled differently.

Tires and Wheels

As with other IH tractors of this era, there was no "standard" tire equipment: everything was considered an attachment. However, there was a default set of wheel, rim, and tire equipment was shipped with a tractor in case a dealer forgot to specify the equipment on an order.

Default rear wheels for the Farmalls were 363028R94 with W11x38 rims until May 23, 1956, when the attachment became 363027R93 with 11x38 rims. The tire and tube part number for the Farmall 450 and 450 D was 354789R91,

When farmers needed to pull a regular implement with a Fast-Hitch tractor, a two-pronged flat drawbar could be inserted into the hitch. Fenders were still considered an attachment in most states in the U.S. at this time.

Table 6.11
Variable Tread Front Wheel and Rim Attachment-Farmall

Attachment number	Rim Size	Rim Manufacturer	Wheel
354765R92	4.25KA-16DC	Goodyear Firestone Cleveland Welding Electric Wheel French & Hecht	8284-DB
359028R91	5.50F-20	IH Goodyear Firestone Cleveland Welding Electric Wheel	09504-D
354091R91	5.50E-20DC	Unknown	09504-D
357515R91	W6-24	Unknown	09504-D

357515R91 eliminated April 7-1958.

Table 6.12
Front Wheels and Rim Attachment-International

Attachment number	Rim Size	Rim Manufacturer	Wheel
363228R91	5.50-F18DC	Goodyear Firestone Cleveland Welding Electric Wheel	363229R12

Table 6.13
Front Tires and Tubes- Farmalls

Attachment number	Tire Size	Plies	Tread	Used with Rims
354108R91	6.00x16	4	F-2	4.25KAx16DC
360818R91	6.00x16	4	F-1	4.25KAx16DC
354109R91	6.00x16	6	F-2	4.25KAx16DC
354787R92	9.00x10	8	F-3	Single Front
370800R91	7.50x16	10	I-1	Single Front
354785R91*	6.50x16	6	F-2	4.25KAx16DC
356754R91*	6.00x20	6	I-1	4.50Ex20DC
359031R91*	7.50x20	4	F-2	5.50Fx20
357519R91**	7.50x24	6	I-1 (FTC9300K)	W6-24

*=Hi-Clears **= Hi-Clear and regular Farmalls, tractors with adjustable wide front axle.
357519R91, 360818R92 eliminated 4-7-1958. 355436R91 changed to 370800R91 same date. The changes were to accommodate tire equipment to be used in 460/560 production.

Table 6.14
Front Tires and Tubes-International

Attachment number	Tire Size	Plies	Tread	Used with Rims
357984R91	6.50x18	4	F-2	5.50Fx18DC
355395R91	7.50x18	4	F-2	5.50Fx18DC
355397R92	7.50x18	4	F-1	5.50Fx18DC
355396R91	7.50x18	6	F-2	5.50Fx18DC
369214R91	7.50x18	8	I-1	5.50Fx18DC

Attachment 355396R91 added November 15, 1956. Attachment 369214R91 added July 18, 1957.

Default front wheels for the Farmall 400, 450, 400 D, and 450 D were part 354765R92, which included 4.25KAx16DC rims, while the tire and tube attachment was 354108R91, including 6.00x16 4-ply F2 tread tires. The high-clearance Farmalls got wheel and rim attachment 354091R91, which included 4.50Ex20DC rims, and tire attachment 356754R91, which included 6.00x20 6-ply I-1 tread tires. The Internationals got wheel and rim attachment 363228R91, which included 5.50Fx18 drop-center rims and tire attachment 357984R91, including 6.50x18 4-ply F-2 tread tires.

Front Axles
Standard front axles for Farmall tractors were the two-wheel narrow front; Hi-Clears all got high-arch wide front axles. The Hi-Clear axles were fitted with mounting pads for the auxiliary stay rods that helped support the taller tractor. The Hi-Clear and the International tractors shared the steering knuckle base part, but the knuckles were finished differently. The Hi-Clear knuckles were made in a right and a left, while the International knuckles were interchangeable.

Bolsters for the Hi-Clear tractors were made of gray iron, while the regular Farmalls were malleable iron. The Internationals had a gray iron bolster.

Front axle 48700-DE was used for production; 48700-DB was used for foreign orders and service. The foreign axles (which apparently were riveted directly to the bolster with special rivets) were made available for domestic service in December 1958.

Single front wheels were available as attachments for Farmall 400, 400 D, 450, and 450 D tractors. There were two different attachments: one for use with 7.50x16, 10-ply front tires (wheels were made by both Motor Wheel Corp. and Electric Wheel Co. under part 354195R91) and one for use with 9.00x10, 8-ply tires (wheels made by IH in a male and female half). The front fork was the same for both attachments. Different sheet metal side plates were used for each attachment.

Fuel Tanks
The liquid-fueled Farmall 400, Farmall 400 Hi-Clear, and gasoline/kerosene/distillate W-400s (not diesels) received a new fuel tank, announced in the Farm Equipment Sales Bulletin No. 27, printed in May 1956. The new tanks were reshaped to fill the space between the transmission housing and the belt pulley attachment, and the

using 12x38 6-ply R-1 treads. On May 27, 1957, the high-clearance default tire and tube changed to 354790R1 with 6-ply R-2 tread 13x38 tires—from 363027R91 and 12-inch tires. The Internationals had as a default rear wheel package 367954R91 with 12x30 rear rims, and 354941R91 for the tube and tire package, with wide base type 14x30 6-ply R-1 tread.

air trap was reduced at the top of the tank. By doing so, the tanks gained 3½ gallons of extra fuel capacity. The change occurred on Farmall 400, 38333, and W-400, 3596. The changes were authorized in December 1955 and required revisions in fuel lines and governor control parts routed around the tank.

Seat

Seat 351880R92 was standard on all 400 and 450 Series tractors, and was a plain metal pan seat made by Milsco.

Tilt-back seat brackets were standard on the International tractors. Shock absorber 351750R93 (made by Monroe and Gabriel, differing somewhat in external appearance) was used on the Farmalls, while 361033R91 (Monroe) was used on the Internationals. Lengths were about an inch different.

Deluxe-type upholstered seat attachments were available for all 400 Series tractors until June 18, 1956, when factory attachment 357819R91 was made standard for International W-400, and then carried on for the W-450. Part 0354155R92, factory, and 357518R91, field, attachments, were available for the Farmalls throughout production, and had silver Koroseal coverings.

Detachable seat pads were available as field attachments on all Series tractors. They came in two styles: deluxe (351438R93) and standard 359483R91. The deluxe had a molded foam rubber bottom, while the standard had a square bottom filler of jute felt. Both were covered with silver Koroseal duck (changed to off-white imitation leather in 1963). Separate field attachments were eliminated in June 1955, and thereafter field and factory attachments were carried under one part number.

A tilt-back seat attachment was available for all Farmall 400 and 450 Series tractors. Originally, it was available as factory or field, but was changed to field only June 18, 1956, then became a parts accessory September 25, 1956. IH started manufacturing the tilt-back bracket in May 1955 under license from Monroe Auto Equipment Co., the previous supplier. The tilt-back seat and deluxe-type seat were combined, authorized in June 1956.

Three different seat springs were available as field attachments: one for lightweight operators (354670R2), one for medium-weight operators (354598R2), and one for heavyweight operators (354671R2).

Swinging Seat

The International W-450 Series had available as an attachment a Swinging Seat Suspension Attachment. The attachment became available in December 1956. The attachment allowed the operator to swing the seat to one side of the platform.

Hydraulics

All hydraulic systems were considered an attachment. However, due to the construction of the tractors, some hydraulic equipment had to be fitted to the tractor to assemble even a basic tractor, mainly the hydraulic reservoir. If a customer had second thoughts, the hydraulic system could be completed in the field by adding the parts from

Table 6.15
Rear Wheels and Rim Attachments- Farmall

Attachment number	Rim Size	Rim Manufacturer	Tire Size	Wheel
357522R93	8.00Tx44	Goodyear Firestone (Attached Clamp)	12-44	365345R1
363027R93	W12-38	Goodyear Firestone (Double Beaded)	13x38 12x38	363014R4
363028R91	W11-38	Goodyear Firestone (Double Beaded)	12x38 11x38	363014R4
365945R91	W14x38	Goodyear Firestone (Double Beaded)	15.5x38	363014R4

357522R93 eliminated April 7, 1958. 365391R91 added December 20, 1955. 365945R91 added May 23, 1956.

Table 6.16
Power Adjusted Rear Wheels and Rim Attachment-Farmall

Attachment number	Rim Size	Rim Manufacturer	Tire Size	Wheel
366099R91	W11x38	Motor Wheel Firestone	11x38 12x38	366101R1
366100R91	W12x38	Motor Wheel Firestone	12x38 13x38	366101R1
367126R91	W14Lx38	Motor Wheel Firestone	15.5x38	366101R1

Table 6.17
Rear Wheels and Rim Attachment-International

Attachment number	Rim Size	Rim Manufacturer	Tire Size	Wheel
354927R91	W12-26	Goodyear Firestone Cleveland Welding	14-26	8299-DA
365391R91	DW20-26	French & Hecht Electric Wheel (Disc Type)	18-26	363290R91
354931R91	DW14-30	Goodyear Cleveland Welding	14-30 15-30	9520-DC
367964R91	W12-30	Firestone (Double Beaded)	13-30 14-30	9520-DC
367965R91	W14-30	Firestone Electric Wheel (Double Beaded)	14-30 15-30	9520-DC
354930R91	DW12-30	Goodyear Firestone Cleveland Welding	13-30 14-30	9520-DC

354930R91 eliminated December 17, 1957. 365391R91 added December 20, 1955. 367965R91 added December 17, 1956.

Table 6.18
Rear Tires and Tubes-Farmall

Attachment number	Tire Size	Plies	Tread	Used with Rims
354129R91	11x38	6	R-1	W11-38
354130R91	11x38	6	R-2	W11-38
354789R91	12x38	6	R-1	W11-38 or W12-38
354913R91	13x38	6	R-1	W12-38
357526R91	12x44	6	R-2	8.00T-44
354131R91	12x38	6	R-2	W11-38 or W12-38
354790R91	13-38	6	R-2	W12-38
365942R91	15.5-38	6	R-1	W14-38

365942R91 added May 23, 1956.
W11-38 cut out of attachment 354789R91 May 23, 1956. 354913R91 eliminated same date.
357526R91 eliminated April 7, 1958.

Table 6.19
Rear Tires and Tubes- International

Attachment number	Tire Size	Plies	Tread	Used with Rims
354940R91	13x30	6	R-1	DW12-30
354941R91	14x30	6	R-1	DW12-30 or DW14-30
354942R91	15x30	6	R-2	DW14-30
355991R91	15x30	6	R-1	DW14-30
354938R91	14x26	6	R-3	DW12-26
357807R91	18x26	8	R-3	DW20-26
355401R91	18x26	8	R-2	DW20-26

Attachments 357807R91, 355401R91 added December 20, 1955.
354940R91 eliminated April 7, 1958.

stock—it wasn't an attachment, so the customer probably had to pay full parts price!

Single-, double-, and triple-valve Hydra Touch systems were available on all tractors as attachments. Control handles were diecast aluminum. Thompson and Pesco produced the pumps, with diesels getting different pumps than carbureted tractors. Hoses were produced by a variety of vendors over the years, including Anchor Coupling, Flex-O-Tube Co., Eastman Mfg., Resistoflex Corp., and General.

Fast Hitch was available on Farmall 400, 400 D, 450, and 450 D with double- or triple-valve Hydra Touch systems for attachment factory or field. All tractors shipped from the factory with Fast Hitch did not have the Fast Hitch mounted, but was shipped with the tractor.

The Farmall 450 and 450 D had available Fast Hitch with Traction Control, with articulation, and the system could be put on the Farmall 400 and 400 D in the field. (The hitch was originally ready to be put onto Farmall 400 Series tractors in May 1956, but was held up so that the hitch could be used as a selling point for the 450s.) Originally, the

The 120-A cotton picker could be converted from cotton-picker use by taking off the picker, reversing the operator's position, installing a regular-clearance front axle, and removing the drop axles. The drop axles were bolted to flanges on the axle carries, the gearboxes were taken off, and the wheels slid onto the exposed end of the axle. Randy Leffingwell

Cotton pickers tend to be pretty large, considering they're mounted on a specially modified tractor to increase clearance, then keep going up from there! Not many people collect pickers because they're difficult to transport, but they do draw attention at shows. Randy Leffingwell

Far left: The pickers used revolving spindles mounted on drums to pick the cotton, which was taken off the spindle and then blown into the basket. Pickers made several passes throughout the picking season to harvest the cotton as the bolls opened. Randy Leffingwell Left: The shifter and steering extensions can be seen here. The regular instruments on the panel still worked. Randy Leffingwell

The Farmall 450 gas was a revised Farmall 400 with the two-tone color scheme, larger engine, and weight-transfer Fast Hitch. The seat on this one is a later aftermarket replacement.

The radiator grille was painted white on these tractors. The standard front axle was the narrow front, although wide front axles were becoming increasingly popular during this period. IH mufflers were made of aluminized steel, and originals had the IH logo and part number stamped onto them.

Pilot Guide (which indicated hitch position so the operator could control the hitch) control-depth-indicator cable ran under the tractor, but complaints of the cable getting snagged forced a relocation of the cable to a route were it was less vulnerable, and a relocation kit was made available for the tractors. The kit was authorized in March 1957.

Hydraulic remote-control adapter packages were available for the Farmalls and Internationals, complete with breakaway couplers, for tractors not fitted with them at the factory. Auxiliary junction blocks were also available as field attachments for tractors with two- or three-valve Hydra Touch systems.

Attachments

Auxiliary Stay-Rod Attachment

An auxiliary stay-rod attachment for stabilizing the front axle was available factory or field on 400 and 450 series Hi-Clear tractors.

Auxiliary Junction Block

Farmall 400 Series tractors with double- or triple-valve Hydra Touch systems had available for a short time a field attachment to add an auxiliary junction block. This was eliminated November 23, 1954.

The background on the nameplate was a decal, not painted. Distributor caps and both ends of the coils were left unpainted by IH. The branding-iron decal was from later IH history but looks pretty cool.

Below left: The 450s went to turn-key ignition with a starting button. Note the safety-type fuel cap on the left side of the hood, seen to this side of the air pre-cleaner.

Above: On the right side of the instrument panel is a shiny button. IH used these to cover holes in the panel that weren't used for accessories. At the top is the panel's light.

Cultivator Stay-Rod Anchor Bracket

This part was available originally as a factory or field attachment, but the field attachment was converted to a parts accessory status on September 27, 1956, and remained available throughout production on Farmall 400, 400 D, 450, and 450 D tractors.

Air-Intake Pipe Extension

Air-intake pipe extensions were available for all 400 and 450 tractors, factory or field, the bracing was different for the diesels.

Pre-Cleaners and Pre-Screeners

Two different pre-cleaners were available for factory or field attachment. The first was the traditional Mason jar–type air pre-cleaner (357869R91) made by Donaldson. Also available was the clear, plastic detachable sleeve–type pre-cleaner with the large, cylindrical plastic body (359314R91) made by United Specialties. In November 1955, new Donaldson pre-cleaners were authorized, eliminating the Mason jar–type, and using the plastic body instead. The part numbers changed from part 59096-D (Mason jar–type) to 365261R91 (2 ?-

This tractor has the newer, 23-degree angled lugs on the rear wheels. They are much easier to find, and are somewhat more efficient than the older, correct-for-this-vintage 45-degree-lug rear tires.

This tractor is fitted with a three-valve Hydra-Touch system. The throttle control is on the quadrant on the steering shaft. The button on the IPTO lever (seen here on the lower left) was bare metal painted red in original production, but a rubber tip is probably a pretty good idea for operation.

inch plastic receptacle). However, the Mason jar–type continued to be available for some time from dealership stocks.

Pre-screeners were available factory or field (field was replaced by designation as a parts accessory), and were produced by Donaldson and United Specialties.

Fuel Cap Gauge

Fuel cap gauges were available as parts accessories. Gauge 365495R92 was used with 18-gallon tanks, and gauge 365497R92 was available for 21½ gallon tanks. The author recommends using the Navistar-offered safety gas caps if you're using gasoline as a fuel.

Increased-Capacity Fuel-Tank Conversion Package

Farmall 400 carbureted-series tractors built with the original smaller fuel tank could be retrofitted with the later, larger fuel tanks with a conversion package from IH. The package differed by whether the tractor was fitted with a two- or three-valve Hydra Touch system, and originally by whether the tractor had a kerosene or distillate attachment in the Farmalls. The International W-400 also had an increased-capacity fuel tank package. The packages became available in June 1956.

A Farmall 450 gas (left) and a 450 diesel (right). When you start examining an individual model of tractor in depth, dozens of variations (if not more) can be found. It's possible that IH may have produced two identical tractors at some point in its history, but they were probably ordered that way by a customer!

Radiator Shutters

Radiator shutters were available as an attachment factory, field, or as a parts accessory for all 400 and 450 Series tractors. The shutter itself was the same on all tractors, but the control linkages varied by tractor and by what hydraulic equipment was on each tractor. IH eventually moved to design systems that could be used on tractors that had either a single-valve Hydra Touch system, or two or three valves. Tractors fitted with shutters also received a radiator-shutter, control-rod operating direction decal (1000696R1 English, 1000763R1 Spanish, 1000815R1 French).

Mufflers and Spark Arrester

A round muffler was available, attachment 357392R92. Two different mufflers were available with this attachment, 357388R92 (manufacturer unknown to the author), and 360078R91 (Donaldson). Mufflers had the IH symbol and part number embossed on the shell, and were made of aluminized steel.

An exhaust-pipe extension was available that was basically just a larger straight pipe. The extension was available for all 400 and 450 Series tractors.

The IH two-piece spark arrester (attachment 51579-D) was available, factory or field, on all 400 Series tractors.

Mufflers on production tractors (and IH parts sold at dealerships) came stamped with the IH logo and part numbers, even through they were produced by outside manufacturers.

99

The left side of the Farmall 450 diesel. Series number plates are on the left clutch housings on 400 and 450 series tractors, they're on the right side on 300 and 350 series tractors.

Oil gauges and hour meters differed between the diesel and gas tractors. All diesels used 12-volt electrical gauges, while some gas tractors still used 6-volt systems.

An exhaust extension was available for factory or field application on all 400 and 450 series tractors, involving the steel tubing extension and a clamp.

Service Meter
A Veeder-Root hour meter was available as an attachment factory or field for all Farmall and International tractors. Slightly different meter drive gears and driven gears were used for the gasoline/LPG- and diesel-burning tractors. Tractors without Hydra Touch hydraulics had a magneto driveshaft that drove the meter.

Steering Wheel
A 20-inch-diameter steering wheel was available as an attachment on Farmall 400 tractors, but was canceled November 1, 1956.

Belt Pulley
Belt pulleys were available for all 400 and 450 Series tractors. Three belt pulleys were available as attachments, factory or field, all manufactured by Browning: an 11-inch diameter by 7½-inch face (358397R93), a 12-inch diameter by 8½-inch face (358597R92), and a 13-inch diameter by 8½-inch face (358598R92).

Hydraulic Power Steering
Power steering for the Farmall 400 was announced in May 1956. The power steering could be either factory installed or applied in the field. The pressure for operating the system was supplied by the Hydra Touch pump. A flow divider valve was used to provide power to the steering and to implements simultaneously.

Power steering was available as an attachment factory or field for International W-400, W-450, W-450 D, and W-400 D tractors, both for tractors without Hydra Touch systems, and for those with Hydra Touch (one, two, or three valves). Attachments differed by whether the tractor was carbureted, or LPG- or diesel-fueled. IH had at least three different potential suppliers for each hose.

Tractors fitted with power steering got an aluminum nameplate stating such and a mono-

Above: As farms got bigger in the 1950s, demand increased for bigger tractors. This, in turn, led to bigger farms, but also to the removal of fence lines and other barriers. Left: The LP tank had to be kept fairly close to the original width of the tractor to maintain forward visibility, which was crucial for cultivating. This tractor has the 23-degree angle modern lugs on the rear tires.

grammed steering wheel cap (made by Gits Molding Corp., Cruver Mfg. Co., and Breyer Molding Co).

Fenders
Fenders were an attachment on the Farmall 400 and 450 series tractors, originally factory or field. After June 21, 1957, the field attachment was eliminated and the fenders could be ordered from parts stocks.

Safety Lamp
The remote safety lamp packages were available for all tractors in 6- and 12-volt systems. The lamps were made by Guide Lamp. When not mounted on an implement, the lamps were stored on a bracket mounted on the tractor.

12-Volt Conversion Package
IH offered a conversion package to refit 400 and 450 Series tractors with 6-volt electrical systems to 12-volt systems. The package was authorized May 28, 1957, and was available until October 1959.

Pneumatic Tire Pumps
Pneumatic pumps by Enginair and Schrader were available factory or field. diesel tractors with the

The carburetor and regulator were unique to the LP tractors and were purchased from Ensign. The carburetor was certainly painted red in production, the regulator probably was as well.

This tank still has the tank manufacturer's data plate. Care should be taken to preserve this valuable plate. IH didn't manufacture its own LP tanks; they came from established specialty manufacturers.

attachment needed a carbureted tractor to actually run the pump!

Wheel Weights

Weights were available factory or field for use with pneumatic tired tractors only. Rear wheel weights for the Farmalls were 6818DA, weighing 145 pounds each, and used up to three on a side. International W-400 and W-450 Series tractors used weight 9114-D, weighing 145 pounds each, up to three a side.

Front weights for Farmall 400, 400 D, 450, and 450 D were 6788-D, weighing 42½ pounds each, and could be used up to two on a side. Farmall 400 HC, 400 DHC, 450 HC, 450 DHC, and International W-400, W-400 D, W-450, and W-450 D, used front wheel weights 9042-D, weighing 73 pounds each, up to two on a side.

Split-type rear wheel weights were available for both the Farmall 400 and 450 Series tractors. International 400 and 450 Series tractors that had disc-type rear wheel and rim 365391R91 also could be fitted with the split weights. All split-type weights could be applied in three layers per side. Split weight 366109R1, weighing 75 pounds each, was used on the Farmalls, while 366645R1, weighing 75 pounds, each was used on the Internationals.

Front-Frame Channel Weights

Left-hand-and right-hand-side frame channel weights were available for Farmall 400, 400 D, 450, and 450 D tractors with Fast Hitch. The weights weighed about 125 pounds each and were made of gray iron.

Front Wheel and Rim

For Farmall tractors originally fitted with a single front wheel, a wheel and rim accessory was used with wide-tread front axle attachment 36040R92

LP tractors had 12-volt electrical systems. The motors' higher compression required a little more oomph to turn over. The lights and electrical gauges are also 12-volt. The shifter knob would have been painted red in production.

for farmers wishing to convert the tractor from the single wheel. The wheel and rim attachment was 367773R91, and included hub 8284-DB and rim 354246R91 (4.25KAx16DC).

Front Steel Wheels
Front steel wheels were available for Farmall 400, 400 D, 450, and 450 D tractors, and the International tractors. The Farmalls got 22½-inch diameter, 4-inch face steel wheels 6738DC (disc-type, cast by IH), which were also one attachment for the Internationals. The other was a 28-inch diameter, 4½-inch-face front wheel, 66657D; a French & Hecht spoke-type.

Rear Steel Wheels
International 400 and 450 tractors had wide face–type steel wheels available as an attachment. These were originally available as a factory or field attachment, but were made a parts accessory September 27, 1956. Fifth gear was locked out on these tractors, while decal 1000708 (gearshift) was used instead of the five-speed shifting decal. On field or parts accessory applications, the rear frame cover had to be drilled for the fifth-gear lockout screw, while factory applications used a predrilled cover (part 362531R21).

Front Skid Rings
Tractors with front steel wheels had available as an attachment factory or field 2½-inch tall skid rings.

There were two sizes: one ring for the 22½-inch-diameter wheels used on Farmall and International tractors, and the 28-inch-diameter wheels used only on the International 400 and 450 Series tractors. Two different parts were used for the smaller wheels: 53014-DA or 67893-DA, purchased from French & Hecht or made by IH. Part 66659-D, used on the larger wheels, was purchased from French & Hecht.

Rear Steel Over-Tire
International tractors with 42-inch-diameter steel wheels and the 5-inch lugs had available factory or field (field-only after September 25, 1956) a 3½-inch-wide over-tire. Bolts were square-headed, with hex nuts.

The LP-gas plate probably had red paint between the lettering in production.

The International 400, and later the W-400, were "standard," non-row-crop tractors that found homes on farms that didn't require Farmalls. While commonly thought of as "Wheatland" tractors, in fact they found homes on dairy and other farms as well. The "W" wasn't added to the tractor's designation until several months after production started.

The 400 used smaller-diameter tires than the Farmall, which lowered the tractor (and the center of gravity). Usual tires were 30 inch, versus the standard Farmall size of 38. Of course, optional sizes were available for each. Rim color depended on rim manufacturer, but was some variation of silver, gray, or galvanized on these tractors.

Spade Lugs

Three different spade lugs were available for steel wheels, all three 3 inches wide. Spade lug 57459-DA was 5 inches tall, 59181-D was 6 inches tall, and 66075-D was 4 inches tall. The 6-inch-tall lugs were used only on the French & Hecht heavy-duty rear wheels used on the Farmalls. The 5-inch lugs were used on all International 400 and 450 tractors with steel, and on Farmalls with the French & Hecht heavy-duty wheels. The 4-inch lug was used on Farmalls with the non-French & Hecht wheels.

Steel Extension Tire

Steel extension tires were available for Farmall and International 400, 450, 400 D, and 450 D tractors, factory or field. The Farmall versions were 51 inches in diameter, while the International extensions were 43 inches in diameter.

Rear-Wheel-Hub Parts Accessory

Starting in May 1956, IH offered a parts accessory to change the rear wheel hub to accommodate disc-type wheels with 16x26 rims and 18x26 tires on Farmall 400, 400 D, 450, and 450 D tractors. The hub used was 365956R3; it was used with a

A rear view of the International 400. These tractors were fitted with the live PTO, but did not have Fast Hitch available—they were designed for use with trailed implements. Of course, hydraulic remotes were optional for implement cylinders.

Above: The operator controls for the throttles and hydraulics were somewhat different from the Farmall's due to the different seat position, steering shaft locations, etc. If these parts need replacement, they can be hard to come by due to low production numbers. Right: PTO shields tend to get battered and abused. They get stepped on, hit with shafts, hit with wagon tongues, etc.

Above: This tractor is fitted with the optional swinging drawbar. One sign of how hard a tractor has been used is drawbar condition—look for stretched out hitch holes or welds indicating repairs. Of course, replacement parts will probably look new. Above right: The parts show a crank-style shutter control, but the decal placement charts and the lever on the tractor show a quadrant-style lever for the radiator shutter control on the International 400s. Below right: An original 12-volt decal from a diesel W-400.

rear-wheel-hub flange, 365957R2 (R1 before November 1958). The accessory was used with disc wheels 559035R91, 554779R91, and 363289R92.

Rear Wheel Changeover Packages

For Farmall 400 Series tractors with serial numbers 5180 and below, two changeover packages were available. Package 364477R92 was a rear wheel with 11x38 double-beaded rim (rim was 363022R1, supplied by Goodyear, Firestone, Cleveland Welding Co., French & Hecht, and Electric Wheel Co). Package 364775R92 was a 12x38 wheel and rim (rim was 363023R1, sup-plied by the same companies). The rear wheel for both packages was 363014R4 (R3 before May 27, 1957).

Tubeless-Tire Trial Lot

A total of 1,000 Farmall 400 and 400 D tractors were shipped to the IH sales districts of Kankakee, Peoria, Broadview, and Springfield, Illinois, with Goodyear 6-ply, 12x38 tubeless tires mounted on 11x38 rims treated to withstand calcium chloride. The decision authorizing the trial lot was dated May 12, 1955.

Adjustable Wide-Tread Front Axle

Adjustable wide-tread front axles were available as an attachment, factory or field, for Farmall 400, 400 D, 450, and 450 D tractors, and were used with any type of front wheels listed for dual application. The wide front axles could not be used with HM-639 cultivators or beet harvesters.

Rear Axle Extensions

Rear axle extensions were available as a field attachment (field accessory after September 25, 1956) for Farmall 400, 400 D, 450, and 450 D tractors. The attachment consisted of a steel axle shaft and a cast steel clamp to join the extension with the rear axle. Extension 353327R2 (R3 after July 1957, with the addition of some induction hardening) was used with double spline axles. Extension 368914R1 was the single spline version (R2 after July 1957, with induction hardening). Two spline axles had to be installed to use the 353327R2 attachment, according to a note added July 23, 1957.

Wide-Tread Rear Axle

Extra-wide rear axles were available for Farmall 400, 400 D, 450, and 450 D tractors. The extra-length axles increased the maximum tread from 94 inches to 106 inches. The attachment was available either factory or field. After June 11, 1957, axle 0368916R1 was used in production (changed to R2 July 19, 1957) while 52147-DC continued as a service part (DD after July 19, 1957, after additional induction hardening was added).

Power Takeoffs

Two types of PTOs were available on all 400 and 450 Series tractors: transmission-driven and Independent. Transmission-driven PTOs were the older-style PTOs with power being supplied through the clutch; when the clutch was disengaged, so was the PTO. These PTOs are not often

A lot of International 400s were used in Great Plains and Western wheatfields, where large fields made high horsepower desirable, and rolling terrain made a low center of gravity desirable. Shields and fenders keep dust and grit (from dry soils) from blowing on the operator.

This International 400 has an air pre-screener that is fairly unusual today; everyone wants the more ornate pre-cleaners. Pre-screeners kept large chunks out of the air cleaner.

seen in 400 and 450 Series tractors but are found and are available. PTOs were attachments and available factory or field, as long as the correct transmission attachment was used. When the PTO was in use, a drawbar extension plate had to be used with tractors equipped with a standard or Fast Hitch.

The Independent PTOs had power from the engine (running through the clutch, but not controlled through the clutch). The Independent PTO shaft ran inside a transmission shaft to the rear of the tractor, where a rear-control gearbox engaged and disengaged it. With this system, the operator could stop the tractor without stopping the PTO, a great help with haying, combining, and other farm operations.

There were actually two different types of Independent PTOs: one for tractors with Torque Amplifier and one for tractors without, which differed slightly. A seasonal disconnect removed

Above: The wide front axle of this International is non-adjustable. There was no need for crop clearance, so it is fairly close to the ground and designed for strength. Right: The toolbox is located under the fuel tank. The rods running up top are for radiator shutters. Below: The International 400's live hydraulic pump was mounted on the ignition drive like the Farmalls. In original production, the distributor cap and plug wires were left unpainted along with the caps on the coil; the coil body itself was red.

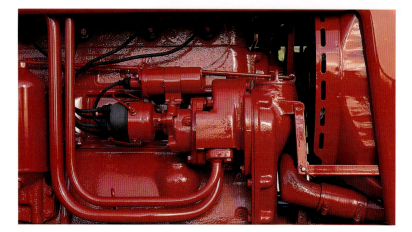

power to the Independent PTO for periods of long disuse, but this was eliminated as standard on December 15, 1955. At the same time, the seasonal Independent PTO disconnect decal that was part of the package was eliminated. The decal was number 1001661R1 (English, 1001668R1 Spanish, 1001669R1 French). In December 1956, the seasonal disconnect was offered as a factory attachment on the 450 series, and as a field package for the 400 and 450 Series tractors with Independent PTO. In February 1957, the field package was eliminated.

Knob 359108R1 was used to activate the Independent PTO and was made of steel by IH. The pin used to fasten the knob was eliminated December 22, 1955; the revised knob became R2.

Electrall
A new and somewhat unusual attachment became available for the Farmall and International W-400 tractors in early 1955. The Electrall was a small generator that could be mounted on the clutch housing and driven by a belt off of the Independent PTO drive gear of the tractor, or driven by the PTO and mounted on a small wagon or on a Fast Hitch mounting. The electricity could be used to either back up the normal power grid or run small power tools around the farm, far from an outlet.

IH examined several other uses for the Electrall. One major push was to use the Electrall as a replacement for the rotating shaft of the PTO, instead running cables between implement and tractor. A combine and hay baler were marketed briefly using these ideas. Another use had the Electrall on the hay baler mounted behind the combine, which was driven by a normal PTO shaft. The baler was attached to the Electrall by a long extension cord. Straw from the combine was fed directly into the baler, eliminating several operations.

The Electrall was an interesting idea, but it failed to sell well. Either the generator was too small (to provide backup power for a farm), too inefficient (the belt pulley drive had a low mechanical efficiency), or too big (running a big tractor engine was an expensive way to drive small power tools). IH closed out the line in the late 1950s.

The attachments were for 400s or 450s with Independent PTO drives, or with Independent PTO-type Torque Amplifiers. The attachments were available factory, with field attachments available up through 1959. The generator units were suspended from turnbuckle and clevis arrangements, with a belt running from the trac-

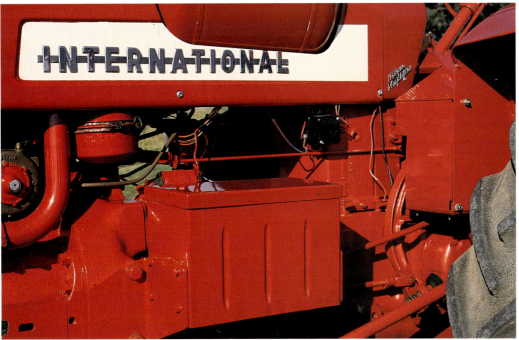

Above: Like the Farmall 450, the International 450 was also produced in an LPG model. All of the W-400-family tractors were replaced by 450s with white painted radiators, white background decals, 281 cubic-inch motors, and other improvements. Left: The battery box hides the Torque Amplifier from view. The LPG tractors had 12-volt electrical systems. These battery boxes can be hard to replace because acid eats the bottoms out.

tor to the generator. The units were hung off the right side of the tractor. IH also sold different configurations as separate implements.

Drawbar
Regular fixed non-swinging drawbars were considered an attachment (factory and field) on all Farmall 400 and 450 Series tractors. The drawbars could be added in the field by taking parts from dealership stock.

High-Clearance Drawbar
The "Quick-Attachable, High-Hitch Heavy-Duty Drawbar" attachment was available as a factory or field attachment for the Farmall 400 and 450 Series high-clearance tractors.

Today a lot of International W-series tractors are coming into the Midwest and other areas that didn't see many of these tractors originally. Their unusual nature, when compared to the far-more-numerous Farmalls, is attractive to today's collectors.

Swinging Drawbar
On the Farmall tractors, the swinging design depended on whether or not the tractor was fitted with a Fast Hitch. On the Internationals, the original swinging drawbar was a non-roller-type (attachment number 360395R91), but was changed to a roller-type (attachment number 360395R91) February 2, 1956. Parts could be ordered from parts stocks to convert the non-roller drawbar to a roller-type drawbar. The rework procedure was described in a service bulletin that included a diagram (FSK-7007). The parts manual also showed the parts necessary for the conversion.

Drawbar Extension Plates
Drawbar extension-plate attachments were available for Farmall 400 and 400 Ds; both with Fast Hitch (plate 363445R1) and, after February 1, 1955, with fixed drawbar (plate 66426-D). These plates were later supplied from parts stocks.

Drawbar Brace
Auxiliary drawbar braces were available for Farmall 400 and 450 Series tractors with fixed drawbars. The packages were authorized in February 1957, with the Hi-Clear tractor packages eliminated in July 1957, if in fact they ever existed.

Fast Hitch Drawbar Stabilizer
Farmalls equipped with Fast Hitch with Traction Control and the Fast Hitch Drawbar had available a drawbar stabilizer as a parts accessory after November 1956.

Magnetos
The trusty IH H-4 magneto was available as a factory attachment for all 400 and 450 Series tractors. Electrical system delete attachments included a magneto and a starting crank offered for Farmall 400 HCs, 450 HCs, 400 DHCs, and 450 DHCs.

Tachometer
After May 1956, tachometers were available for all 400 Series tractors not fitted with magnetos. The tachometers were available as factory attachments or parts accessories for field attachment. The tachometers read in miles per hour (based on rpm), with 12x38 tires for the Farmall versions, and 14x30 tractors for the International W tractors.

This tractor has the optional wheel weights installed. In production, an LP-gas decal would be located under the tank data plate.

Front Power-Takeoff Pulley Attachment
A front power-takeoff pulley attachment for the W-450 was authorized December 3, 1957. The pulley could be applied factory or field.

Cotton Picker Conversions
Just as with the Farmall 300 Series tractors, 400 and 450 Series Farmalls (except high-clearance tractors) were converted for use as cotton picker power. The conversions fell into two main cotton picker versions: high-drum and low- drum. The low-drum cotton pickers had reverse operation and other differences. The high-drum had the flanged rear axle and drop housings and a taller front wheel, giving extra ground clearance for the taller cotton picker drum used with taller cotton varieties.

All cotton picker tractors were designed so that they could be converted back and forth from cotton picker configuration to normal. Today, ironically, many collectors display high-drum cotton picker tractors halfway between configuration, with forward operating configuration but with the drop housings and taller axle installed. After the cotton pickers wore out, the cotton pickers were converted to spray rigs and operated this way.

Field attachment 364481R91 was designed "To convert tractors previously equipped with cotton picker mounting attachments (high or low drum) back to normal farm operation." Bolster 6719-DBX and front axle 8700-DH were used, along with hub and rim assembly 355036R93.

Cotton picker conversions were fitted with a special spark arrester to avoid setting the cotton on fire, as well as to direct the exhaust away from the cotton basket. At first, these were an attachment, but they were made mandatory after February 28, 1955. Originally, 400 Series cotton pickers had an "orchard-type exhaust," but this was eliminated February 28, 1955.

Decals
These are the decals commonly found on 400 Series tractors. Some decals were used in conjunction with attachments or special equipment, so see the individual section for those decals.
2750017R1 decal, hood sheet nameplate background, L.H. (450 Series only)
2750018R1 decal, hood sheet nameplate background, R.H. (450 Series only)
1001734 decal, lighting switch

The visibility problems caused by a larger fuel tank weren't as pronounced with the W-450 as they were with the Farmalls. The radiator grill was, of course, painted white.

1000680R2 decal, gasoline (English, 1000770R2 Spanish, 1000822 French) diesel- and kerosene/distillate-burning tractors only

1000984R2 decal, warning, pressure cooling (English, 1000985R2 Spanish, 1000986R2 French)

1000686R5 decal, warning power takeoff (English, 1000772R5 Spanish, 1000824 French)

1000704R4 decal, warning, brake (English, 1000767R4 Spanish, 1000819R3 French)

1000813R1 decal, 12-volt electrical system (English, 1000836R1 Spanish, 1000830R1 French)

1001182R1 decal, warning, drawbar and front pull hook (English, 1001184R1 Spanish, 1001185 French)

1001013 decal, instructions, oil filter (English, 1001021R5 Spanish, 1001103 French), (carbureted only)

1001501 decal, instructions, lubrication oil filter (English, 1001502 Spanish, 1000503R1 French), (diesel only)

1000922R4 instructions, lubrication oil filter, auxiliary filter (English, 1000923R4 Spanish, 1001203R1 French), (diesel only)

1001336R2 decal, instructions, final-fuel oil filter (English, 1001337R2 Spanish, 1001338R2 French), (diesel only)

1000889R4 decal, injection-pump lubrication (English, 1000890R4 Spanish, 1000891R4 French), (diesel only)

1001181R3 decal, instruction, air cleaner (Donaldson), (English, 1001205R3 Spanish, 1001206R3 French)

1001289R1 decal, instruction, air cleaner (United Specialties), (English, 1001290R1 Spanish, 1001291R1 French)

1000967R2 decal, instructions, water trap (English, 1000968R2 Spanish)

This tractor has the later, fuller-coverage fenders with a square crown. The tractor's tires kick up quite a bit of dust in the field, and the fenders help minimize grit in the operator's eyes caused by the lower seating position.

1000707R1 decal, gearshift

1001591R1 decal, patent marking

1001725 decal, caution, 6-volt electrical system (English, 1001726 Spanish, 1001727 French), eliminated (carbureted only)

1001722 decal, caution, 12-volt electrical system (English, 1001723R1 Spanish, 1001724R1 French), (diesel and LPG only)

1001718R1 decal "Made in the UNITED STATES of AMERICA" all non-U.S. orders, applied next to serial number plate

Changes

1000922R4 Instructions, lubrication oil filter, auxiliary filter (English, 1000923R4 Spanish, 1001203R1 French), changed to 1001701R1 (English, 1001702R1 Spanish, 1001703R1 French), (diesel only) January 12, 1955

1001591R2 decal, patent marking changed to R2 October 18, 1956, addition or deletion of numbers

1001725 decal, caution, 6-volt electrical system (English, 1001726 Spanish, 1001727 French), eliminated April 13, 1957, all tractors used 12-volt after that

1001701R1 (English, 1001702R1 Spanish, 1001703R1 French) changed to R2 September 25, 1957

1001336R2 decal, instructions, final fuel oil filter (English, 1001337R2 Spanish, 1001338R2 French), (diesel only) changed to R3 September 25, 1957, (changes in filter part number)

Chapter Seven
International 600, International 650

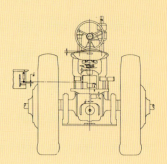

The International 600, introduced in 1956, was the replacement for the McCormick Super-9 series tractors. It was intended for heavy tillage, non-row-crop applications. It was produced at Melrose Park, Illinois, for less than a year. This is the gasoline-powered version.

The 600s were not authorized for production until August 31, 1956, over two years after authorization of other 100 Series tractors. In early planning, the new tractor was known as the W-600, but the W was eliminated in February 1956, before production. The new tractors featured several improvements over the Super-9 Series tractors, including an increase in the steering ratio and the height of the steering wheel, a positive engagement starter to handle the larger engines, new operators platforms and rear fenders, and a hydraulic-type swinging seat. Styling was revamped with a new radiator grille, hood, and instrument panel. The throttle control lever was now lubricated, a weather cap was fitted for the exhaust pipe, and new, larger lights were used.

The 600 saw new C-350 gas engines released, recognition that the tractors could handle more horsepower through the transmissions and rear ends. A new D-350 diesel was also available that had an adjustable jet venturi for the gasoline starting system to help with cold starting.

The 600s were the very first International Harvester tractor models with hydraulic power steering available from the start of production. The system used many parts available for Super WD-

and WDR-9 special attachments, but with new regulators, valve assemblies and covers, and new hoses. A "Power Steering" nameplate was used when the attachment was used on a tractor. The power steering attachment could be fitted either at the factory or in the field.

The International 600 had some problems. The new 650—introduced in 1956 as part of the new line of IH tractors—had several upgrades, unlike most of IH's American product changeovers of this time. A new C-350 gasoline engine and a new D-350 diesel engine increased horsepower. The brake locking mechanisms were revised and new clutch pedals and operating rods were used to reduce clutch pedal effort.

A touch-up of the styling was in order, as well. IH had seen other company's equipment going to a two-tone paint job, most noticeably at the 1956 Pennsylvania Road Show, a sales convention for road building equipment. IH decided to follow suit, creating a white decal to go under the ID nameplates on the hood, while the radiator grille was painted white. The IH hood emblem background was changed to red instead of black.

These tractors featured the styling that IH had been using on its smaller tractors since late 1954, combined with a platform and fenders designed to protect the operator from wind-blown grit. The platform also allowed the operator to stand during long hours of operation.

Table 7.1
Kinds and Codes

Kind 836-International 650D, Domestic Sales- Code 1; Kind 836-International 650, Domestic Sales, Code 2; Kind 836-International 650, Domestic Sales, Code 3-LPG; Kind 558-International 600D/650D Code 1; Kind 558-International 600D/650D Code 2

836 Code 1	836 Code 2	836 Code 3	558 Code 1	558 Code 2	Part Number	Description
x	x	x			355389R91	Rear Wheel & 5.50Fx18DC Rim
x	x	x			355395R91	7.50x18 4 ply F-1 Tread tire & Tube
x	x	x	x		355392R91	rear wheel & DW12-34 rim
x	x	x	x		355403R91	14-34 6 ply wide base R-1 tread rear tire & tube
x	x	x	x		264132R91	Exhaust muffler attachment
x	x	x	x		365565R91	Cigarette Lighter attachment
x	x	x	x		365630R92	Tachometer Attachment
x	x	x	x		357819R92	Deluxe type upholstered seat attachment
			x		355390R91	Pneumatic wheel 5.50F x 18DC rim
			x		355397R91	7.5 x 18 4-ply front tire F-1 Tread
				x	365707R91	Hydraulic Power Steering
				x	365643R91	Hydra-Touch single valve system
		x			366843R91	LPG Burning Attachment

The export tractors had the same equipment as the domestic kinds and codes, but also included oil filter element packages for all tractors, and fuel filter element packages for Diesels.

Table 7.2
Engine Specifications

	Piston	Cylinder Head	Compression Ratio	Gasket	Manifold
Gasoline	365882R1	8320-DB	6.12	52795-DD	8321-DB
LPG	366856R1	8320-DB	8.25	52795-DD	366920R1 (Intake) 366921R1 (Exhaust)
Gas (Service parts after 1959)	365882R1	8320-DC	6.08	52795-DD	8321-DB
LPG (Service Parts after 1959)	366856R1	8320-DC	8.15	52795-DD	366920R1 (Intake) 366921R1 (Exhaust)

The flip-back seat allowed the operator to stand while operating the tractor, as well as keeping the seat dry during rain or heavy dew conditions. Note that the rear lamp is painted the correct red.

Engine
Starting cranks were supplied through the parts system for domestic customers. Cranks were standard for tractors shipped export.

Gas
The 350-ci engine used on the carbureted versions of the tractor had a new combustion chamber. It was designed for high turbulence to improve combustion and to guard against lead deposits, a growing problem with higher-octane gasoline.

Carburetor
An International E-12 was used on gasoline engines.

Valves
Exhaust valves on all 600 Series engines were fitted with valve rotators to improve heat distribution and to reduce deposits. Molybdenum valve-seat inserts improved seat life.

Fuel pump
A fuel pump with a built-in sediment bulb was used.

Manifold
A one-piece manifold was used. A short exhaust pipe with a weather cap was used.

LPG
LPG engines were only available on 650 Series tractors. The LPG-burning engine required many changes to both the engine and chassis. Cylinder heads were the same, although the valve seat inserts were different. Carburetor assembly 366922R91 was used in the LPG tractors.

LPG Manifold
A two-piece manifold was used. The intake manifold had an elbow fitted for the idle line, which fed LPG into the manifold directly for idling.

Diesel
The diesel versions used the traditional start-on-gas system that IH had used since 1933.

Diesel Carburetor
An International carburetor intended solely for starting the tractor was used. The carburetor drain trough, located on the bottom of the carburetor, was used to shift any leakage from the carburetor away from the magneto. The jet in the carburetor was adjustable from the operator's seat.

Electrical
Several different types of electrical systems were available in the 600 and 650 Series tractors. The simplest system was a magneto ignition, without any other equipment. A hand crank would be used to start; possible with the diesel because of the low compression start on the gas system. Magnetos could also be used with a battery starting and lighting system. The International H-4 was made in a carbureted and a diesel version, having different rotors and impulse assemblies. Carbureted tractors had a special kill switch to ground the magneto, turning off the tractor. The diesel tractor had a switch built into the intake manifold cover (on the right side of the intake manifold) that killed the magneto when the engine switched over from gas to diesel.

Distributor-type battery ignitions were much more common in production. All 600 and 650 tractors used a 12-volt electrical system. The distributor was driven from the same drive as the magneto and was horizontal with a coil clipped to the top of the unit. At least three different distributors were used, which were identified by a letter on the base of the distributor housing.

The generators for both carbureted and diesel tractors were identical; however, the carbureted tractors had a two-piece belt pulley, while the diesel tractors had one-piece pulleys. The carbureted and diesel versions also used different mounting brackets and braces. Voltage regulators and mountings were identical across all 600 and 650 tractors. Delco-Remy manufactured the starters.

Instrument panel lights were used on tractors with lighting systems. The lights used a cover with a hole cut out to aim light at the instrument panel. Twelve-volt light bulbs were used within a lens inside the cover. The light assembly was made by Guide Lamp on the 600s and early 650s, but in April 1957, an optional lamp assembly made by Dietz was authorized.

Fuses on the 600 were originally 10 amps, but went to the 15 amps fuses either very late in 600 production or when the switch was made to the 650. Ignition switches were made by three suppliers under one part number: United Specialties, H.A. Douglas Mfg. Co., and Briggs and Stratton. The starting buttons were also changed at the time of the 600/650 switchover, with 255357R91 on the 600 replaced by 366316R91 or 366317R91 on the 650. The buttons were shaped slightly different.

The two optional headlights were 365818R91 (Guide Lamp) and 365512R91 (Dietz). Both were sealed beams in teardrop-type housings. Rear lighting was the combination rear light and taillight. The rear light housing had a light that was white in color, used for illuminating the implement. Another small light had a red jewel over the light bulb. Lamp clamps were made in steel (350920R1) or malleable iron (6780-D) versions. The light assemblies for the headlamps and combination taillamps changed slightly several times in production of the 650. The rear lamps were available for factory application or in the field as a spare part.

Batteries and battery boxes were different between the carbureted and diesel versions. The diesel used two boxes, side by side, mounted on the left side of the tractor. The carbureted version used one battery box with one 12-volt battery. The batteries were Globe Union or Auto-Lite. The diesel battery boxes had covers with individual rubber hold downs, while the carbureted boxes had a frame-type hold down that fit on the edges of the top of the battery and under the cover itself. Both styles of battery boxes were supported by a shaped flat plate that attached to the frame of the tractor on the left side.

Chassis
Clutch
A 12-inch clutch made by Rockford or IH were used. The clutch compartment cover was made

This photo should answer the question, "Where does the starting crank go?" The oil filter instructions are also seen here.

Spark plug wires were not painted, and neither were distributor caps.

The starter is painted red, with the data plate masked off. Don Corrie, who owns these tractors, prefers his carburetors black, although on original production, they were painted red with the rest of the tractor. Don has a wonderful collection of 600- and 650-series tractors (including all tractors in this chapter) and is considered a leading restorer of IH tractors.

The wide, non-adjustable front axles on the International 600 have little crop clearance. These tractors were most frequently used for the wheat and small-grain farms of the Great Plains and the West, where they usually encountered only stubble. But these tractors did find other uses.

IH tended to start its serial number ranges with the number 501. Here is International 600 diesel serial number 501. Belt pulleys of this era were originally steel and painted all red. However, over time, other pulleys can be found on the tractors as farmers changed sizes to get a particular belt speed.

The diesels needed more battery power to crank them over (although technically, they could still be hand cranked), so IH put two batteries on the 600s. The smaller box in front is the correctly located tool box.

The 600s had either plain or swinging drawbars—they never had Fast Hitch or Independent Power Take Offs. The roomy platform with wide fenders, along with the oval-shaped instrument panel can be seen here.

in sheet steel on the 600 and 650, as opposed to earlier cast iron.

The 600 family tractors and early 650 (up through serial number 797) family tractors used two different clutch pedals, 8876-DB (purchased from a vendor) or 357567R1 (IH made of malleable iron). In October 1956, a change was authorized to use clutch pedal 367130R1, purchased outside the company. The new clutches were used on 650 serial numbers 798 and up. At the same time, the grease fitting was changed from a ⅛-inch fitting (Alemite 1610 or Lincoln 5000) to a ³⁄₁₆-inch fitting (Alemite #1957).

Transmission

Gearshift rubber boot 61064-D was used throughout production, and was a carryover from the previous Super Series tractors. Part 53717-DA was used as the hand lever knob throughout production of the Supers, 600s, and 650s. The gearshift lever on 600s was shorter than the production lever on the previous tractors to clear the new instrument panel.

Radiator

The radiator assembly was the same for all the 600/650 tractors with the exception of the radia-

tor outlet hoses (the diesel hoses were a little shorter) and the outlet pipe (the LPG pipe was different). The fan housings were different between the carbureted and the diesel tractors.

A cooling improvement package was released in 1959 for all -9 Series, 600, and 650 Series diesel tractors. The package was composed of a radiator surge tank that mounted in place of the gasoline starting tank. The gasoline starting tank was then

The diesels had to have extensive fuel filtering systems in order to prevent damage from grit getting into the high-pressure, tight-tolerance injectors and injection pump. Each part of the several-stage filtering system had its own maintenance requirements and instruction decal. See the chapter on decals and paint for more information. In addition, the pump had a decal as well.

The International 650 gas was the follow-on tractor to the 600 gas, featuring the new two-tone color scheme. The radiator grill is painted white. See the chapter on decals and paint for precise information.

relocated to the left side of the engine just behind the radiator. A caution decal stating "Caution, To Avoid Scalding, Always remove surge tank cap before radiator cap" was located on the left side of the grille, with the right edge of the decal lined up with the centerline of the radiator cap. There was a considerable amount of plumbing involved with installing the package.

Air Cleaners

On the 600 D and 650 D, both Donaldson and United Specialties air cleaners were used. However, on the 650, only Donaldsons were used. The

air cleaner decals came in English, Spanish, and French versions. LPG tractors only used an 8-inch type-A Donaldson air cleaner.

Sheet Metal

The 600 introduced a new platform, fenders, and instrument panel. The bottom of the platform was diamond-treaded sheet metal. The crown-type over-the-wheel fenders were attached to both the platform floor and the cowl shields that closed off the area between the front of the fenders and the body of the tractor, which was a Wheatland-type arrangement.

The initial platform was not too successful, and platform service package 367129R91 was designed for all 600 Series tractors and 650 Series 501–797. New platform 365576R11 was used on serial 798 and up.

Fuel tank cap 23995-DC was used throughout production, and was carried over from the Super -9 Series. On diesel and kerosene/distillate versions, a small gasoline tank was used for starting, located under the right side of the hood.

The 600 originally had a center radiator grille molding and two side radiator grille moldings. The elimination of the center grille molding was authorized September 12, 1956, and was effective on the start of 650 production. Since the same grille was used as service parts on the -9 and 600 Series tractors, replacement parts also lacked the holes in the grille for the moldings.

The hood sheets differed by fuel and destination, mainly in holes drilled to accept the nameplate moldings because different names were used outside North America. The diesels had additional holes for the starting gas tank, as did the kerosene-distillate tractors.

During the change from the 600 to the 650, the number plate changed. The "McCormick" nameplate that was used in some foreign markets changed from 362317R1 to 366685R1 September 19, 1956, and involved changing the background of the plate from red to white to match the white background decals.

Radiator Cap

Original radiator caps on 600 tractors were 67448-D. On April 20, 1956, the change to parts 361705R91 (Stant) or 362323R91 (A.C. Spark Plug Co.) was made. The older caps were the "frying pan-type," while newer caps were the "lobed-type."

On January 23, 1957, the radiator cap part numbers were again changed, this time to 361705R91, which was used exclusively. Three different company radiator caps were handled under this number: Stant Mfg. Co., A.C. Spark Plug Co., and Eaton Mfg. Co.

Fenders

Fenders on 650 tractors were extensively revised, authorized in September 1956, to increase tire coverage and improve appearance. There was an optional fender support plate.

There weren't a lot of changes from the 600 to the 650. The 650 continued in production until 1958, the last of the tractors based on the W-9 design. It was replaced by the 660, which was a higher-horsepower version of the 560, itself a six-cylinder continuation of the Farmall M.

The left engine side of the 650 gas. The carburetor is the correct red, and the starter is correctly detailed. The white under the nameplate is not paint—IH put a decal under the nameplate to avoid using two colors on the paint line. The radiator grill was painted separately from the tractor.

Instrument Panel and Controls

The 600s and 650s were fitted with an oval-shaped instrument panel that mounted just over the steering wheel shaft. A cigarette lighter was optional, and could also be supplied through the spare parts system or as an accessory. The lower center of the instrument panel had a large space for the optional tachometer. On tractors without the tach, the sheet metal "knock out" was left intact.

Diesel tractors with a starting and lighting system were also fitted with a choke rod with a knob at the end. Diesel tractors were also fitted with a carburetor fuel adjusting rod for the starting carburetor with a knob at the end. Gas tractors had choke rods with the wire loops at their ends. LPG tractors had cable-within-sleeve choke controls with knobs at their ends.

The control panel on carbureted and LPG tractors differed 365629R1, (gas) vs. 366933R1 (LPG). LPG tractors also used different temperature indicators (366937R91 vs. 365628R91.)

The governor control handle used on 600 D tractors was decreased 3 inches, authorized July 23, 1956, to clear the new swinging seat. The change also affected service parts used on earlier tractors.

Serial Number Plate

The 600 family tractors used serial plate 365546R1, while 650s used 366693R91; the major difference being the model number.

Toolbox

A steel toolbox was regular equipment. The toolbox was located in front of the battery box.

Seats

The seat mounting was the hydraulic-type seat, with a fore and aft adjustment. The seat also swung to either side and locked. A plain pan seat (351880 R92) was standard. Two seat pads were available for the plain seat as spare parts: 351438R93 (the deluxe version) with rubber material and 359483R91 with jute material. Coverings were gray Koroseal-coated cotton duck. The standard seats had a tilt-back bracket so the operator could stand while operating the tractor.

A deluxe seat was available as a factory attachment or as a parts accessory. When installed at the factory, part number 0354155R92 was used. Some seats used in production may have had a small tag saying "foam rubber padding" on them. Coverings were gray Koroseal.

Three different conical springs were available. The spring for lightweight operators, 354671R2, had three identifying notches. The spring for medium-weight operators, 352598R2, had four identifying notches. The spring for heavyweight operators, 354670 R2, had five identifying notches. The lightweight- and heavyweight springs were probably installed either by the dealership or operator and ordered through the spare part system.

Seats used shock absorber 361033R91. The swinging-seat locking pin knob was a $7/8$-inch O.D.

The 650 was available in an LP-gas-powered version. In the 1950s, LP gas was inexpensive in many areas where the 650s were used. Given the amount of fuel the big tractor burned, it was a good idea for IH to offer this LPG version.

x .065 plain flat washer that was part of the pin. The trip rope clips were still mounted on these tractors.

Brake Pedals

On the 600 and 650 tractors, IH could use two different designs of brake pedals. Left pedals 8877-DB were purchased from a vendor, while 357565R1 were cast in malleable iron by IH. On the right pedal 8878-DB was from a vendor, while the IH cast malleable iron pedal was 357566R11. The vendor-produced pedals, fabricated from various steel parts by Kingston Products Corp., also had been used in -9 production. The brake operating lever also came in two versions: 8869-D made of malleable iron and 350156R1 made of steel.

Bull Gears

IH designated two different bull gear hubs to be used in the 600 and 650 tractors (carried over from the SWD-9 family). One hub, 058415-D, was made of forged steel, while the other hub, 0357652R1, was made of gray iron. The different designs were in case of material shortage or high costs.

Fixed Drawbar

The fixed drawbar was standard equipment on the 600/650 tractor. The drawbar came with a drawbar extension plate for tractors equipped with PTO. This brought the hitch dimensions to the ASAE standard so that other makes of machinery and standardized IH machinery could be used.

Wheels and Tires

IH officially considered all wheels and tires as extra-cost attachments. However, when the salesman forgot to specify which wheels and tires the customer wanted, IH did have a default set to ship with the tractor, and the customer was charged accordingly. The default front equipment was a 5.5Fx18 DC rim (IH part #355389R91) and 7.5x18 4-ply, F-2 tread front tire with inner tube. The default rears were DW12x34 (IH #355392R91) rims and 14x34 6-ply wide base R1 tread rear tires with inner tube.

The 600s and 650s had available live hydraulic pumps, seen here as the gray piece in front of the ignition. In production, the pump and hydraulic hoses were painted red.

The LPG system requires the use of a flow regulator (due to high-pressure fuel) in addition to the carburetor. This special equipment was provided by Ensign. This tractor has a power-steering system; the hydraulic rams are at the bottom of the photo.

When fitted with power steering, the 650 has an identification plate on the steering wheel in case the operator couldn't figure it out!

Steel front wheels were available for these tractors. Steel wheel attachment 355406R91 (30-inch diameter, 6-inch face, wheels) or 355407R91 (34-inch diameter, 6-inch face, wheels 69996DX), were available either at the factory or as a field attachment. The 34-inch attachment had been used on the rice field-bound Super WDR-9s, so it may be that the larger front wheels were also put on 600s and 650s bound for soft fields as well, as per customer order.

A 2½-inch skid ring was available with the 34-inch wheel, and a 1½- and a 2½-inch skid ring were available with the 30-inch diameter wheel. An extension wheel for steel fronts, attachment 70047-D was available (the tire was 70048DA).

Most if not all steel wheels and parts were purchased from French & Hecht.

Steel rear wheels were available from the factory in a 48-inch (attachment 361515R91) and a 54-inch (attachment 361516R91) wheel diameter. A number of interesting things happened when the wheels were put on the tractor (they were also available as a field attachment). A different gearshift decal was used (1000708R1) and fifth gear was locked out by a screw. The wheels were purchased whole while the hub was purchased as a rough casting from French & Hecht and then machined by IH. The drawbar on the 48-inch field-attachment tractors was re-drilled. The 54-inch wheel tractors from the factory also had a different drawbar. The 54-inch attachment was used previously in the Super WDR-9 (Rice) Series; 600 and 650 tractors ordered by customers with soft fields probably ordered the larger equipment.

The rear steel wheels also had 6-inch-wide extension tire attachments. The 48-inch steel tire was 62186-D and the 54-inch tire was 70043-D. The steel tires were held on by 16 attaching plates (62184-DA was the actual 48-inch extension tire and 70045-DA was the 54-inch wheel extension).

Available for the 48-inch wheel were 5-inch-high by 3¼-inch-wide lugs; 6-inch-high by 4-inch-wide lugs were available for the 54-inch wheel.

Front Wheels and Rims

For the 5.50Fx18DC rim, two wheels and four different rims were available. Either wheel 8713-

Table 7.3
Rear Wheel and Rim Attachments

Attachment number	Rear Wheel	Rim Size	Type of Rim	Tire Size	Plies	Tread
359761R91	8888-DB	W16-34	Goodyear	15-34	6	R-1, R-2
255358R91	255367R1	DW16-26DC	Goodyear, Firestone, or Motor Wheel	18-26	8	R-1, R-2, R-3 Heavy Duty Rear Axle only
355392R91	8888-DB	DW12-34	Goodyear, Firestone, or Cleveland Welding	14-34	6	R-1
355393R91	8888-DB	DW14-34	Goodyear, Firestone, Cleveland Welding Co	15-34	6	R-2
355394R91	8888-DB	DW14-34	Goodyear, Cleveland Welding	15-34	6	R-1, R-2
363286R91	363289R92 or 354779R91	DW16-26	French & Hecht (disc)	18-26	8	R-1, R-2, R-3
363287R91	363290R91 Sheet 1 Or 363290R91 Sheet 2	DW20-26	French & Hecht (disc)	18-26	8	R-1, R-2, R-3
365759R91	255367R1	DW-20-26D (Heavy Duty)	Firestone	18-26	8	R-1, R-2, R-3

The French & Hecht were disc type combination rear wheels/rims with hubs, while the others had large wheels with rims.

DB or 350954R1 was used. Rims were all carried under part number 351219R92, but there were four different manufacturers: Goodyear, Firestone, Cleveland Welding, and Electric Wheel. On these rims went 7.5x18 tires that were either 4- or 6-ply with the F-2-type tread or 4-ply on the F-1 tread.

Hydraulics
The 600 and 650 family had the Hydra Touch hydraulic system for use with remote cylinders. The system also provided power to the power steering system. One or two valve systems were available. A mounting pad on the front frame of the tractor was available for an auxiliary hydraulic pump required for certain industrial equipment. The pump was driven from the front of the engine crankshaft. A round reservoir tank was strapped onto a bracket. Both single- and double-valve systems could be added in the field (parts supplied from spare parts stock) and the single-valve system could also be changed to a double-valve system in the field. The control levers were die cast.

Attachments
Air Cleaner Extension Pipe
Air-intake pipe extensions were available for both carbureted and diesel engines. The braces differed between the carbureted and diesels.

Pre-Cleaners/Pre-Screeners
A collector-type pre-cleaner was available with the plastic bowl. This attachment was available for factory order or for field attachment as a spare part.

Two different types of pre-screeners were available: a two-piece with an exterior steel shell with large round holes, which featured an interior screen, or a one-piece with a screen exterior. The one-piece pre-screener was part 367725R91 made

Table 7.4
Rear Tire Attachments

Attachment number	Tire Size and Tread	Plies	Used with Rims
355400R91	18-26(R-1)	8	DW 16-26 DC and DW 20-26DC
355401R91	18-26(R-2)	8	16-26DC & DW20-26DC
357807R91	18-26(R-3)	8	DW 16-26DC & DW 20-26DC (after Feb 12, 1956)
355403R91	14-34(R-1)	6	DW 12-34
355404R91	15-34(R-2)	6	DW 14-34 & DW 16-34
355776R91	15-34(R-1)	6	DW 14-34 & DW 16-34

The instrument box/panel for the 650 (and 600) tractors included a shielded light at the top, while the black knob on the right is actually the fuse. The tachometer (gauge at the bottom) used engine RPMs with a scale to show theoretical ground speeds in different gears.

The hydraulic valve handles were probably painted red in production. This tractor has a single valve system, most likely because it was only used with trailing implements having one ram.

125

The big seller in the 650 line was, of course, the diesel, mainly for its economical use of cheap fuel. The 650 diesel was one of the last examples of IH's start-on-gas diesel engines that used a carburetor, ignition, and a complex cylinder head with compression release to start the engine on gasoline, then switch it to diesel once it warmed up.

by United Specialties, first authorized November 20, 1956, and available from the factory or in the field as a spare part. Part number 364564R91 was the two-piece pre-screener made by Donaldson and was available throughout 600 and 650 production either from the factory or for field attachment as a spare part.

Mufflers and Spark Arresters

Weather Caps were standard on the 600 and 650 family tractors. The part number was 281182R91 for most of the production, but very early 600s may have used cap 365685R91. The base level exhaust for the 600 and 650 carbureted and LPG engines was exhaust pipe 58095-DA, which was a steel pipe. On the diesels, the base exhaust pipe had a flange cast into the bottom and was cast in gray iron. The part number of the diesel pipe was 6995-DB.

Mufflers on the 600 and 650 family were either 54127-D or 264132R91. These mufflers were available factory or field. Mufflers were made of aluminized steel, and were stamped with the IH logo and part number.

Spark arresters were available factory or field. The arresters had a cast iron top and bottom piece.

Service Meter

Two different service meter packages were used on 600 and 650 family tractors. The meter itself was the same in the packages. Carbureted tractors needed an ignition unit driveshaft that the diesels did not (carbureted tractors with Hydra Touch also did not need the extra shaft).

Tachometer

The Tachometer was available from the factory or as a field attachment. Tractors with magneto ignitions could not be fitted with tachometers as no provision was made inside the magneto to operate the tach driveshaft. The tachometer registered revolutions per minute, and indicated ground speed in miles per hour in all forward gears. The miles per hour on the dial face were calculated to be accurate with 14x34, 6-ply, R-1 tread rear tires. The tachometer itself was purchased from Stewart Warner. The drive gear for the tachometer was plastic.

Over-Center Hand Clutch

A 12-inch-diameter, over-center hand clutch was available for factory and field applications. The clutch had a long hand lever, which snapped the clutch over center to take it out of engagement, without having to put constant pressure on the lever. It was used where farmers wanted to operate the tractor standing up, including rice and Wheatland farming, where many of these tractors were used. The clutch was available on all 600 and 650 (as well as the earlier Super W-9s) family tractors. Two different clutch disks were used alternately, 70099-D (Rockford) or 258210R92 (IH). When the hand clutch was used, the righthand side of the front platform was modified to attach the lever.

Radiator Shutter Controls

Shutters were available for both gasoline and diesel tractors. The front control rod was different between the diesel and the gasoline tractors. When fitted, the radiator shutter control decal was applied. The English decal part number was 1000696R1, Spanish was 1000763R1, and French was 1000815R1.

Decelerators

Decelerators were foot pedals located on the right side of the platform, that when stepped on, decelerated the tractor. These were often used in rice country for going over levees, but may also have found use in other areas where the tractors were operated from a standing position. The 600 and 650 carbureted tractors used a common decelerator package that was available for factory or field application. The foot pedal was connected to the throttle assembly near the platform.

The diesel decelerators were more complicated. The pedal actuated a linkage that went all the way forward to the injection pump control lever. The diesel decelerators were available factory or field.

The right side of the 650 diesel gets pretty busy with the gas starting part of the engine, as well as the live hydraulic pump and hoses. The small, inclined piece over the distributor is a drip shield that steers gasoline overflowing from the carburetor above it away from the distributor—an important piece of safety equipment.

Power Steering

Tractors fitted with the Hydra Touch hydraulic system had a power steering system available that used the existing hydraulic pump and reservoir, adding a regulator valve and safety flow-control valve to the Hydra Touch system, as well as a control arm and the hydraulic steering cylinder.

On tractors without the Hydra Touch system, a separate hydraulic pump powered by a belt that ran the water pump and fan was used. The Hydra Touch reservoir was used, as well as much of the rest of the regular power steering setup. The power steering system for tractors equipped with Hydra Touch was available for factory or field application, and included power steering nameplate 366215R1, which was a steering wheel insert. It had bright aluminum borders and letters, a black background on the top half, and a red background on the bottom half. The system for tractors without a Hydra Touch hydraulic system was available for field application only.

Hoses were made by several different manufacturers, depending on the hose. Manufacturers included Anchor Coupling, Flex-o-Tube, Eastman, Resistoflex, and General.

Belt Pulley

A variety of pulleys were available. The standard pulley was a 14-inch diameter, 8½-inch face pulley, part number 357340R91 made by Browning Mfg. Co. Another IH-made pulley with the same dimensions, 356945R91, was available in case IH could not get pulleys supplied from Browning.

Brownings were plain metal faces, as were IHs. Two other pulleys, one a 16-inch diameter (357811R91, made by Browning) and one with a 16½-inch diameter (356944R91, IH built for emergency use) were also available. The pulley control knob, 56158-D, was die cast.

Power Takeoff

The 600 and 650 tractors used a transmission-driven PTO that was not "live." The PTO was ASAE standard as far as diameter and length. The PTO shifter rod changed at 650 serial number 837. The 365733R1 was straight while

A rear view of the 650 diesel. The operator's position is a lot more comfortable (in the author's opinion) than on the Farmalls, because the 650 diesels were intended for long hours of operation in big fields.

Injection pumps can be expensive to repair, and have often been replaced with pumps of a different make. Pumps also have their own serial numbers, so in most cases the information can be checked to see when the pump was produced by IH.

366170R1, the later part, was bent at an angle. The PTO attachment was available factory or field. The field attachment used a different PTO shifter rod, 70273-D.

Front Power Takeoff
IH had available for the 650 D tractor a front power-takeoff pulley attachment. This was available for factory application (IH part number 370016R91) or field application (370017R91).

Pneumatic Tire Pumps
Two different pumps were used in the 600 and 650 tractors, Schrader and Engineair. The Engineair pump was manufactured by the G.H. Meiser Co. The pumps could be used with a carbureted engine but not with diesel.

Mud Scraper
Mud scrapers were available factory or field for tractors with rear steel wheels.

Wheel Weights
Wheel weights were only used on tractors with pneumatic tires. For tractors with disc-type rear wheels, split rear weights were used 366645R1 that were about 75 pounds for each half. The weights could be attached in second or third sets.

Tractors with 26-inch rear air tires and cast iron wheels used weights 8077-D. The weight could be applied up to three sets, and weighed about 137 pounds. For 34-inch pneumatic tires, wheel weight 8887-D could be used in up to three sets, with each weight weighing about 140 pounds.

Available front wheel weights weighing 73 pounds apiece (part 9042-D) were used in singles or pairs on each side.

Watchdog Fuel Filter (IH Fuel Filter Only)
The Watchdog fuel filter attachment was available for the International 650 D tractors fitted with the IH fuel pump. The attachment was available for factory or field attachment. The filter consisted of a short, round canister located above the injection pump, with a hose leading from the regular fuel filters into the bottom of the canister. The hose leading to the pump ran out of the top of the canister. The attachment was handled under number 278675R91.

Swinging Drawbar
A swinging drawbar was available on all 600 and 650 tractors. The drawbar was the same as found on the Super -9 Series tractors and was certainly available for field application through the parts merchandising program, but probably from the factory as well.

Safety Lamp and Breakaway Socket
A safety lamp package was available for the 600 and 650 tractors. All lamps were 12-volt. The lamp assembly was made by Guide Lamp under their part number 897105. Tractors with the lamp package had to have the breakaway connector socket that mounted under the rear of the platform. The cable to the breakaway socket ran from the rear light down the inside of the right-hand fender. A wire clip was located about halfway between the light and the fender, and another was at the base. When ordered from the factory or for field application, a simple electrical cable was used; when service parts were ordered, harness 365476R91 was used.

The electrical breakaway socket itself was considered a separate option, available from the factory or as a field attachment. IH used sockets from two different manufacturers (Cole Hersee Co. and Joseph Pollack Co.) under the same part number of 362002R92. Two wiring harnesses were used, one from the rear lamp (that replaced the regular rear lamp harness) and one leading to the lighting switch. A "Caution, 12 Volt electrical system" decal was used when the socket was installed.

Gas Caps
The original standard cap was flat, part 23995-DC. A gauge-type fuel cap made by Rochester

was available through the dealerships as a parts accessory. However, a tall-type fuel cap with improved ventilation is now highly recommended for safety purposes, modern fuels being somewhat more variable in quality than fuel in the 1950s. The new caps are available from dealerships, but if your tractor has an original cap, the new caps can be obtained for free from International Truck and Engine.

Heavy-Duty Rear-Axle Attachment
A heavy-duty rear-axle attachment was available for a while on 600 and 650 tractors approved on June 27, 1956, during 600 production, but canceled March 11, 1957, during 650 production, due to very low demand (it's unclear as to whether or not any were built). This attachment used the 16x26DC rims and 51304-D wheel weights. The 255368R1 rear wheel and 255366 or 258237R91 rims were used.

600/650 Decals
1000867R1 lighting switch decal
2750030R1 decal hood sheet nameplate background L.H. (650 family only)
1000984R2 warning-pressure cooling (English, 1000985R2 Spanish, 1000986R2 French)
2750031R1 decal, hood sheet nameplate background, R.H. (650 family only)
1000636R5 warning, power takeoff (English, 1000772R5 Spanish, 1000824R4 French)
1000819R3 warning, brake (French)
1000813R1 warning, 12-volt electrical system (English, 1000836R1 Spanish, 1000830R1 French)
1001182R1 warning, drawbar and front pull hook (English, 1001184R1 Spanish, 1001185R1 French)
1001501R1 instructions, oil filter (diesel only) (English, 1001502R1 Spanish, 1001503R1 French)
1001701R2 instructions, oil filter (auxiliary diesel filter)(English, 1001702R2 Spanish, 1001703R2 French)
1001336R3 instructions, fuel-oil final filter (English, 1001337R3 Spanish, 1001338R3 French)
1000889R4 injection pump lubrication (English, 1000890R4 Spanish 1000891R4 French)
1001181R3 air cleaner (gas), (English, 1001205R3 Spanish, 1001206R3 French)
1001289R3 air cleaner (diesel), (English, 1001290 Spanish, 1001291R1 French)
1000967R2 instructions, water trap (diesel only), (English, 1000968R2 Spanish)
1000707R1 gearshift
01001591R2 patent marking
1001718R1 decal "Made in the United States of America" (all export, plus Puerto Rico)
1001014R5 oil filter instructions (gas), (English, 1001022R5 Spanish 1001104R5 French)
1000696R1 radiator shutter control (when used), (English, 1000763R1 Spanish, 1000815R1 French)

Changes
01001591R2 changed from R1 October 18, 1956, patent numbers were added or deleted
On September 25, 1957, the auxiliary fuel filter changed from R1 to R2, and the final fuel filter changed from R2 to R3. Replacement fuel filter element parts numbers were changed

The 650 diesel had a different oil-pressure gauge than the 650 gas, and the hour meter was a little different because of differing engine RPMs.

The 650 diesel used two batteries for the starting system. The smaller box is the tool box. The boxes do rust out from battery acid, and it can be somewhat pricey to buy reproductions, let alone find original boxes.

129

Appendix A
Serial Numbers and Production

Serial Numbers
IH did not produce tractors with differences by model year such as a '55 Chevy. Changes were made sometimes that coincided with a "year," but many changes were made throughout a year of production.

CUB
The Cub was produced from 1947 to the late 1970s/early 1980s. Sheet metal changed in late 1954 and again in 1956 to match the larger tractors introduced at that time.

Yearly Serial Numbers
- 1954: 179412–186440 (new-style Cub production covered by this book starts at 185000)
- 1955: 186441–193657
- 1956: 193658–198230 or 198231
- 1957: 198231–204388
- 1958: 204389–211440 (end of Cubs covered in this book)

Monthly First Serial Numbers—Chassis

	Jan	Feb	Mar	Apr	May	Jun	Jul	Aug	Sep	Oct	Nov	Dec
1954											185001	185647
1955	186441	187307	188136	189080	189948	190819	191737	192283	192377	192531	192802	193134
1956	193658	194218	194983	195748	196345	196697	197025	197262	197570	None	197825	197976
1957	198231	198812	199405	200034	200676	201063	201456	201931	202380	202858	203416	203891
1958	204389	204833	205226	205399	205581	205808*						

*205808 last Hundred Series-style tractor off line

Monthly First Serial Numbers—Engines

	Jan	Feb	Mar	Apr	May	Jun	Jul	Aug	Sep	Oct	Nov	Dec
1954									184293	185035	185574	186189
1955	187052	187876	189914	190412	191706	195589	196629	197377	197629	198105	198672	199680
1956	200633	201476	202370	205829	206480	207223	207198	207868	208263**	-	208026	210382
1957	210686	212535	213326	214122	215083	215648	216293	217089	217884	218604	219330	220027
1958	220802	221423	221977	222302	221857	222807*						

*last Hundred Series-style off line
**there is a discrepancy between the September 1956 and November 1956 Engine Serial Lists

Production Numbers

	Produced for U.S.	Produced for Canada	Produced for Foreign	Total
1955	7160	282	354	7801
1956	4635	172	213	5020
1957	5429	109	53	5591
1958	2270	64	59	2393

three modified Cubs produced 1956, 1957, 1958

Care should be taken when cleaning the tractor to not remove the paint from the serial number tag. Replacements are available, but the original is good to have, especially if the tractor is rare.

CUB LO-BOY

Yearly Serial Numbers
- 1955 501–2554
- 1956 2555–3928
- 1957 3929–6581
- 1958 6582–10566

Monthly First Serial Numbers—Chassis

	Jan	Feb	Mar	Apr	May	Jun	Jul	Aug	Sep	Oct	Nov	Dec
1955							501	789	945	1203	1645	2208
1956	2555	2863	2973	3086	3195	3362	3465	3556	3674	None	3772	3830
1957	3929	4131	4336	4552	4776	4935	5187	5482	5809	6034	6179	6375
1958	6727	6851	6960	7074	7123*							

*last Hundred Series off line; production with different stylings continued well after

Monthly First Serial Numbers—Engines

	Jan	Feb	Mar	Apr	May	Jun	Jul	Aug	Sep	Oct	Nov	Dec
1955							196902	197429	197653	198082	198673	199572
1956	200634	201561	202386	202900	206658	206953	207325	207685	208265	None	208489	210387
1957	210641	212541	213358	214160	215131	215652	216443	217098	217883	218600	219335	220036
1958	220375	221439	221998	222285	222049	222665**						

**last 1958 style off line

Production Numbers

	Produced for U.S.	Produced for Canada	Produced for Foreign	Total
1955	1122	30	1	1153
1956	6654	273	214	7141
1957	2357	37	13	2407
1958	945	0	0	945

FARMALL AND INTERNATIONAL 100 & 100 HC

Yearly Serial Numbers

FARMALL 100 AND 100 HIGH CLEARANCE
- 1954 501—1719
- 1955 1720—12894
- 1956 12895—18940

INTERNATIONAL 100
- 1954 501—503
- 1955 505—574
- 1956 575—635

Monthly First Serial Numbers—Chassis Farmall 100 and Farmall 100 HC

	Jan	Feb	Mar	Apr	May	Jun	Jul	Aug	Sep	Oct	Nov	Dec
1954											501	1070
1955	1720	2440	3141	4281	5274	6295	7389	8371	8723	9321	10356	11635
1956	12895	11419	14966	15819	16549	17181	17688	18077	18550	-		

130

Monthly First Serial Numbers—Engines Farmall 100 and Farmall 100 HC

	Jan	Feb	Mar	Apr	May	Jun	Jul	Aug	Sep	Oct	Nov	Dec
1954											572	1389
1955	3212	3209	5223	8055	11274	12948	14902	16771	17404	18374	20019	22229
1956	24172	26497	28997	30869	31068	31123	33728	34433	35359	-		

Monthly First Serial Numbers—Chassis International 100

	Jan	Feb	Mar	Apr	May	Jun	Jul	Aug	Sep	Oct	Nov	Dec
1954											501	502
1955	504	504	513	521	577	605	612	631	633	635		

Monthly First Serial Numbers—Engines International 100

	Jan	Feb	Mar	Apr	May	Jun	Jul	Aug	Sep	Oct	Nov	Dec
1954											976	1694
1955		4581	5845	7926	11054	12968	15208	16926	18001	No Number		

Production Numbers

FARMALL 100 GAS

	Produced for U.S.	Produced for Canada	Produced for Foreign	Total
1955 Gasoline	7842	644	29	8515
1955 Distillate	375	0	180	555
1955 Kerosene	0	0	115	115
1956 Gasoline	7301	518	88	7907
1956 Distillate	211	0	26	237*
1956 Kerosene	0	0	43	43*

*eight modified F-100s produced in 1955, three in 1956

INTERNATIONAL 100

	Produced for U.S.	Produced for Canada	Produced for Foreign	Total
1955	80	0	0	80

• replaced by Farmall 100s with badging for International

FARMALL 100 HC

	Produced for U.S.	Produced for Canada	Produced for Foreign	Total
1955 Gasoline	396	1	18	415
1955 Distillate	113	0	18	244
1955 Kerosene	0	0	3	3
1956 Gasoline	273	2	10	285
1956 Distillate	61	0	49	110
1956 Kerolene	0	0	0	0

On Cub and Cub Lo-Boy tractors, the serial number tag is on the front steering bolster, seen here under the steering shaft.

MODEL 130

Yearly Serial Numbers

1956 501—1119
1957 1120—8362
1958 8363—10209

Monthly First Serial Numbers—Chassis

	Jan	Feb	Mar	Apr	May	Jun	Jul	Aug	Sep	Oct	Nov	Dec
1956											501	734
1957	1120	1820	2516	3253	4024	4758	5312	5858	6406	6902	7399	7871
1958	8363	8786	9158	9505	9867							

*last tractor 10209 built in May

Monthly First Serial Numbers—Engine

	Jan	Feb	Mar	Apr	May	Jun	Jul	Aug	Sep	Oct	Nov	Dec
1956											36046	36418
1957	37057	39019	40665	42854	44497	46452	48023	48695	50172	50385	52403	53400
1958	54425	55975	56357	57366	57928							

*last engine built 58560

Production Numbers

FARMALL 130

	Produced for U.S.	Produced for Canada	Produced for Foreign	Total
1957 Gasoline	5824	484	35	6343
1957 Distillate	117	0	0	117
1957 Kerosene	0	0	16	16
1958 Gasoline	2452	169	0	2621
1958 Distillate	28	0	0	28
1958 Kerosene	0	0	14	14

*48 modified Farmall 130s produced in 1957, 10 modified Farmall 130s produced in 1958

FARMALL 130 HC

	Produced for U.S.	Produced for Canada	Produced for Foreign	Total
1957 Gasoline	294	1	7	302
1957 Distillate	53	0	67	120
1957 Kerosene	0	0	0	0
1958 Gasoline	115	2	1	118
1958 Distillate	8	0	22	30
1958 Kerosene	0	0	0	0

MODEL 200

Yearly Serial Numbers

1956 501—1031
1957 1032—10903
1958 10904—15698

Monthly First Serial Numbers—Chassis

	Jan	Feb	Mar	Apr	May	Jun	Jul	Aug	Sep	Oct	Nov	Dec
1954											501	745
1955	1032	1654	2510	4979	5832	6712	7636	8477	8676	8996	9549	10232
1956	10904	11600	12686	13749	14489	14815	15073	15267	15504	None		

Monthly First Serial Numbers—Engines

	Jan	Feb	Mar	Apr	May	Jun	Jul	Aug	Sep	Oct	Nov	Dec
1954											663	1416
1955	1331	3870	5548	8004	10774	12894	14410	16407	16951	17166	20011	22004
1956	24697	26588	28537	30756	32258	33277	32865	34716	35009	None		

Production Numbers

Farmall 200	Produced for U.S.	Produced for Canada	Produced for Foreign	Total
1955 Gasoline	7875	671	17	8563
1955 Distillate	255	0	13	268
1955 Kerosene	0	0	215	215
1956 Gasoline	5494	432	21	5947
1956 Distillate	166	0	1	167
1956 Kerosene	1	0	35	36

Model 230

Yearly Serial Numbers

1956501—814
1957815—6826
19586827—7671

Monthly First Serial Numbers—Chassis

	Jan	Feb	Mar	Apr	May	Jun	Jul	Aug	Sep	Oct	Nov	Dec
1956											501	619
1957	815	1293	1785	2292	2802	3276	3772	4318	4866	5359	5856	6329
1958	6827	7166	7435	7521	7591	7671						

*47 modified serial numbered tractors made in either April or May 1958

Monthly First Serial Numbers—Engines

	Jan	Feb	Mar	Apr	May	Jun	Jul	Aug	Sep	Oct	Nov	Dec
1956											36001	36427
1957	36878	39054	40514	42809	44657	46486	47711	49124	50130	51197	52504	51953
1958	54240	55954	56509	56910	56930							

*last engine built in May

Production Numbers

Farmall 230	Produced for U.S.	Produced for Canada	Produced for Foreign	Total
1957 Gasoline	4861	357	17	5235
1957 Distillate	103	0	0	103
1957 Kerosene	0	0	17	17
1958 Gasoline	1725	28	26	1779
1958 Distillate	24	0	1	25
1958 Kerosene	0	0	12	12

157 modified 230s produced in 1958, 125 in 1957.

Farmall and International 300

Yearly Serial Numbers

Farmall 300, Farmall 300 HC

1954501—3359
19553360—23223
195623224—30508

International 300

1955501—20218
195620219—33664

Monthly First Serial Number—Chassis F300

	Jan	Feb	Mar	Apr	May	Jun	Jul	Aug	Sep	Oct	Nov	Dec
1954											501	1677
1955	3360	5637	8659	10659	12236	13820	15831	17715	18110	18933	20035	21623
1956	23224	24823	26416	27708	28725	29243	29578	29862				

*Last tractor built 30508

Monthly First Serial Numbers—Chassis Farmall 300 HC

Sep	Oct	Nov	Dec		Jan	Feb	Mar	Apr	May	Jun	Jul	Aug
1955	1779?	5046	6131	9972	12071	13877	15875	17590	18028	18555	20053	21538
1956	24002	25524	26544	27988	28891	29375	29608	29866	30258			

*Last tractor built 30474

Monthly First Serial Numbers—Chassis International 300

	Jan	Feb	Mar	Apr	May	Jun	Jul	Aug	Sep	Oct	Nov	Dec
1955			501	2225	4334	6447	9012	11367	11860	13067	14902	17551
1956	20219	22872	25527	27783	29610	30544	31352	31901	32807			

*Last tractor built 33664

Monthly First Serial Number—Engine Farmall 300

	Jan	Feb	Mar	Apr	May	Jun	Jul	Aug	Sep	Oct	Nov	Dec
1954											501	1680
1955	2363	5640	8662	12372	16070	19767	24346	28584	29473	31502	34443	38679
1956	43002	47363	51762	55610	58803	60305	61561	62449	63809			

*Last tractor built 65208

Monthly First Serial Number—Engine Farmall 300 HC

	Jan	Feb	Mar	Apr	May	Jun	Jul	Aug	Sep	Oct	Nov	Dec
1955	1782	5046	6134	10782	15683	19900	24446	28303	29288	30494	34491	39356
1956	45074	49282	52104	56494	59256	60860	61650	62461	63970			

*Last tractor built 65074

Monthly First Serial Number—Engine International 300

	Jan	Feb	Mar	Apr	May	Jun	Jul	Aug	Sep	Oct	Nov	Dec
1955				12373	16069	19768	24345	28585	29472	31503	34441	38680
1956	43003	47359	51758	55606	58802	60306	61559	62460	63806			

*Last tractor built 65210

Production Numbers

Farmall 300	Produced for U.S.	Produced for Canada	Produced for Foreign	Total
1954	4	0	0	4
1955	18500	441	336	19277
1956	9755	312	157	10224

*Three Farmall 300s scrapped at factory in 1956

Farmall 300 HC	Produced for U.S.	Produced for Canada	Produced for Foreign	Total
1955	53	0	12	65
1956	104	0	1	105

*14 Farmall 300s HC shipped to Piper Paine for conversion in 1956

International 300	Produced for U.S.	Produced for Canada	Produced for Foreign	Total
1955	12537	1732	43	14312
1956	16774	2036	54	18864

*12 International 300s Shipped for Piper Paine conversions in 1956
*47 International 300s Shipped to Seaman Gunnison in 1956
*Four International 300s scrapped at factory in 1956.

International 330

Monthly First Serial Numbers—Chassis International 330

	Jan	Feb	Mar	Apr	May	Jun	Jul	Aug	Sep	Oct	Nov	Dec
1957											501	731
1958	1488	2365	3165	4001								

*Last tractor 4763 built April 25, 1958

Production Numbers

International 330	Produced for U.S.	Produced for Canada	Produced for Foreign	Total
1958	3972	290	0	4262

Model 350

Yearly Serial Numbers
Farmall 350s
    1956 501—1029
    1957 1030—14673
    1958 14674—17215

Monthly First Serial Numbers—Chassis Farmall 350

	Jan	Feb	Mar	Apr	May	Jun	Jul	Aug	Sep	Oct	Nov	Dec
1956											501	707
1957	1004	1797	2797	4143	5765	7531	8516	9543	10359	11287	12316	13220
1958	14175	15114	15886	16570								

*Last tractor 17215 built April 25, 1958

Monthly First Serial Numbers—Chassis Farmall 350 HC

	Jan	Feb	Mar	Apr	May	Jun	Jul	Aug	Sep	Oct	Nov	Dec
1956											533	872
1957	1030	2311	3287	4491	None	None	8711	9576	10671	11593	12849	13975
1958	14674	15452	None	16616								

*Last tractor 17150 built April 24, 1958

Monthly First Serial Numbers—Chassis Farmall 350 D

	Jan	Feb	Mar	Apr	May	Jun	Jul	Aug	Sep	Oct	Nov	Dec
1956											581	708
1957	1003	1800	2796	4141	5766	7529	8518	9542	10361	11286	None	13713
1958	14176	15115	15871	16568								

*Last tractor 17149 built April 24, 1958

Monthly First Serial Numbers—Chassis International 350

	Jan	Feb	Mar	Apr	May	Jun	Jul	Aug	Sept	Oct	Nov	Dec
1956											501	1103
1957	1963	3484	5075	6431	7500	8365	9563	10884	12060	13095	13958	14616
1958	15049	15788	16546	17483								

*Last built 17149 built April 24, 1958

Monthly First Serial Numbers—Chassis International 350 D

	Jan	Feb	Mar	Apr	May	Jun	Jul	Aug	Sep	Oct	Nov	Dec
1956											739	1102
1957	1978	3485	5077	6439	7498	8367	9568	10899	12069	13100	None	14832
1958	15051	15791	16547	17489								

*Last built 18274 April 24, 1958

Monthly First Serial Numbers—Engine Farmall 350

	Jan	Feb	Mar	Apr	May	Jun	Jul	Aug	Sep	Oct	Nov	Dec
1956											502	1189
1957	2355	4330	6413	8216	10127	12026	14011	16110	17874	19600	21322	22884
1958	24173	25520	26700	27863								

*Last engine built 28951

Monthly First Serial Numbers—Engine Farmall 350 HC

	Jan	Feb	Mar	Apr	May	Jun	Jul	Aug	Sept	Oct	Nov	Dec
1956											629	1786
1957	2440	5347	7063	8685			14410	18181	18548	20084	22308	23943
1958	24796	26030	27980									

Monthly First Serial Numbers—Engine International 350

	Jan	Feb	Mar	Apr	May	Jun	Jul	Aug	Sept	Oct	Nov	Dec
1956											501	1190
1957	2296	3023	6411	8231	10134	12024	14012	16106	17873	19599	21323	22886
1958	24175	25519	26699	27860								

*Last engine built 28950

Production Numbers

Farmall 350	Produced for U.S.	Produced for Canada	Produced for Foreign	Total
1957	8498	71	82	8651
1958	3324	44	30	3398

Farmall 350 LPG	Produced for U.S.	Produced for Canada	Produced for Foreign	Total
1957	Not Broken Out			
1958	242	0	0	242

FARMALL 350 HC	Produced for U.S.	Produced for Canada	Produced for Foreign	Total
1957	82	0	0	82
1958	69	0	0	69

FARMALL 350 D	Produced for U.S.	Produced for Canada	Produced for Foreign	Total
1957	2909	49	2	2960
1958	1206	17	8	1231

FARMALL 350 DHC	Produced for U.S.	Produced for Canada	Produced for Foreign	Total
1957	11	0	1	12
1958	8	0	1	9

INTERNATIONAL 350	Produced for U.S.	Produced for Canada	Produced for Foreign	Total
1957	9522	761	34	10317
1958	2452	76	6	2534

*Four International 350s were scrapped at the factory in 1957

INTERNATIONAL 350 WHEATLAND	Produced for U.S.	Produced for Canada	Produced for Foreign	Total
1957	10	809		819
1958	no separate #			

INTERNATIONAL 350 D WHEATLAND	Produced for U.S.	Produced for Canada	Produced for Foreign	Total
1957	4	142	0	146
1958	No Separate #			

INTERNATIONAL 350 LPG	Produced for U.S.	Produced for Canada	Produced for Foreign	Total
1957	Not Broken Out			
1958	147	0	0	147

INTERNATIONAL 350 D	Produced for U.S.	Produced for Canada	Produced for Foreign	Total
1957	2102	189	4	2295
1958	670	66	2	738

*Two International 350-D scrapped at factory in 1957

INTERNATIONAL 350 HU	Produced for U.S.	Produced for Canada	Produced for Foreign	Total
1957	Not Broken Out			
1958	871	6	0	877

INTERNATIONAL 350 HU LPG	Produced for U.S.	Produced for Canada	Produced for Foreign	Total
1957	Not Broken Out			
1958	55	0	0	55

INTERNATIONAL 350D HU	Produced for U.S.	Produced for Canada	Produced for Foreign	Total
1957	Not Broken Out			
1958	73	0	0	73

Farmall 300- and 350-series tractors have serial number plates on the right side of the clutch housings, while the 400 and 450s have them on the left side.

MODEL 400

Yearly Serial Numbers
Monthly First Serial Number—Chassis Farmall 400

	Jan	Feb	Mar	Apr	May	Jun	Jul	Aug	Sept	Oct	Nov	Dec
1954											501	2455
1955	4732	7481	10389	13111	15227	18453	20628	22383	22753	23660	25041	27032
1956	29065	30934	32799	34612	36237	37675	38860	39788	40767			

*Last tractor built 41484

Monthly First Serial Number—Chassis Farmall 400 HC

	Jan	Feb	Mar	Apr	May	Jun	Jul	Aug	Sept	Oct	Nov	Dec
1954												2517
1955	2588	6823	10132	11466	15567	18388	20576	22442	22768	23381	24828	26966
1956	29290	31926	33821	35138	36887	37951	38862	40243	41323			

*Last tractor built 41422

Monthly First Serial Number—Chassis Farmall 400 D

	Jan	Feb	Mar	Apr	May	Jun	Jul	Aug	Sept	Oct	Nov	Dec
1954											505	2457
1955	4733	7501	10407	13115	15727	18461	20629	22382	22755	23661	25042	27096
1956	29064	30939	32801	34611	36238	37676	38861	39791	40766			

*Last tractor built 41485

Monthly First Serial Number—Chassis Farmall 400 DHC

	Jan	Feb	Mar	Apr	May	Jun	Jul	Aug	Sept	Oct	Nov	Dec
1954												
1955			8191	12205	12260	18279	20288	22127	22127	22127	22127	26732
1956	29435	31836	32909	35109	36450	37923	39619	None	41081			

*Last tractor built 41081

Monthly First Serial Number—Chassis International 400

	Jan	Feb	Mar	Apr	May	Jun	Jul	Aug	Sept	Oct	Nov	Dec
1954												
1955	510	681	801	1083	1349	1515	1650	1761	1785	1843	1931	2059
1956	2187	2443	2697	2964	3221	3478	3652	3746	3819			

*Last tractor built 3858

Monthly First Serial Number—Chassis International 400 D

	Jan	Feb	Mar	Apr	May	Jun	Jul	Aug	Sept	Oct	Nov	Dec
1954												510
1955		684	803	1084	1348	1514	1648	1762	1786	1844	1934	2042
1956	2192	2442	2702	2965	3220	3476	3651	3745	3820			

*Last tractor built 3857

Monthly First Serial Number—Engine Farmall 400

	Jan	Feb	Mar	Apr	May	Jun	Jul	Aug	Sept	Oct	Nov	Dec
1954											136834	138406
1955	140211	142692	145613	148386	150467	153420	155387	156764	157082	157741	158792	160398
1956	161969	163556	165141	166708	168139	169405	170413	171202	171992			

*Last tractor built 172564

Monthly First Serial Number—Engine Farmall 400 HC

	Jan	Feb	Mar	Apr	May	Jun	Jul	Aug	Sept	Oct	Nov	Dec
1954												138455
1955	138527	142035	145361	146713	150766	153363	155344	156857	157093	157538	159029	160709
1956	162157	164427	165989	167154	168712	169648	170414	171612	172448			

*Last tractor built 172515

Monthly First Serial Number—Engine International 400

	Jan	Feb	Mar	Apr	May	Jun	Jul	Aug	Sept	Oct	Nov	Dec
1954												
1955	137831	142708	145615	148392	150993	153442	155411	156767	157089	157745	158795	160414
1956	17933	163564	165143	166707	168149	169417	170426	171209	171991			

*Last tractor built 172562

Monthly First Serial Number—Engine International 400

	Jan	Feb	Mar	Apr	May	Jun	Jul	Aug	Sept	Oct	Nov	Dec
1954											11783	12313
1955	12892	13394	13460	13632	14018	14612	15188	15786	15974	16245	16847	17485
1956	18031	18516	19089	19616	20107	20545	20840	21130	21460			

Production Numbers

Farmall 400	Produced for U.S.	Produced for Canada	Produced for Foreign	Total
1954	1	0	0	1
1955	19281	135	508	19924
1956	11498	111	272	11881

Farmall 400 HC	Produced for U.S.	Produced for Canada	Produced for Foreign	Total
1955	69	1	16	86
1956	52	0	5	57

10 F-400 HC shipped to Littleford for conversion in 1956
23 F-400 HC shipped to Littleford in 1955

Farmall 400 D	Produced for U.S.	Produced for Canada	Produced for Foreign	Total
1955	3961	43	279	4283
1956	3910	52	308	4270

Five Farmall 400 D shipped to Seaman Gunnison in 1956

Farmall 400 DHC	Produced for U.S.	Produced for Canada	Produced for Foreign	Total
1955	10	0	5	15
1956	12	0	23	35

International 400	Produced for U.S.	Produced for Canada	Produced for Foreign	Total
1955	434	518	16	968
1956	430	630	42	1102

International 400 D	Produced for U.S.	Produced for Canada	Produced for Foreign	Total
1955	180	130	58	368
1956	298	426	106	830

Model 450

Yearly Serial Numbers

Farmall 450
- 1956 501–1773 or 2606 (IH sources disagree)
- 1957 1734–18338 or 2607-22304 (IH Sources Disagree)
- 1958 22305–26067
- 1958 1662—2295

W-450s
- 1956 501–567
- 1957 568—1661

Monthly First Serial Numbers—Chassis Farmall 450

	Jan	Feb	Mar	Apr	May	Jun	Jul	Aug	Sept	Oct	Nov	Dec
1956											501	1009
1957	1734	2911	4182	5512	7025	8674	10403	12391	14085	16085	18839	20245
1958	21871	23231	24299	25328								

Last tractor 26066 built April 25, 1958

Monthly First Serial Number—Chassis Farmall 450 HC

	Jan	Feb	Mar	Apr	May	Jun	Jul	Aug	Sept	Oct	Nov	Dec
1956											596	1470
1957	2058	3322	4247	None	None	9797	None	None	None	16320	19395	20987
1958	21885	23935	none	26025								

Last tractor 26025 built April 24, 1958

Monthly First Serial Number—Chassis Farmall 450-D

	Jan	Feb	Mar	Apr	May	Jun	Jul	Aug	Sept	Oct	Nov	Dec
1956											585	1022
1957	1737	2913	4186	5514	7024	8677	10405	12392	14086	16089	18430	20246
1958	21874	23230	24297	25329								

Last tractor 26067 built April 25 1958

Monthly First Serial Number—Chassis Farmall –450 –DHC

	Jan	Feb	Mar	Apr	May	Jun	Jul	Aug	Sept	Oct	Nov	Dec
1956											None	None
1957	2607	None	None	5992	8082	8686	None	13474	14789	None	None	None
1958	22305	None	24907	25589								

Last tractor 25832 built April 17, 1958

Monthly First Serial Number—Chassis International W-450

	Jan	Feb	Mar	Apr	May	Jun	Jul	Aug	Sept	Oct	Nov	Dec
1956											501	528
1957	568	678	821	974	1111	1237	1322	1414	1498	1549	1624	
1958	1661	1766	1887	2015								

Last tractor 2295 built April 25, 1958

Monthly First Serial Number—Chassis International W-450 D

	Jan	Feb	Mar	Apr	May	Jun	Jul	Aug	Sept	Oct	Nov	Dec
1956											None	529
1957	567	677	820	970	1109	1236	1320	1415	1500	None	1613	1625
1958	1662	1763	1884	2011								

Last tractor 2294 built April 25, 1958

Monthly First Serial Numbers—Engine Farmall 450

	Jan	Feb	Mar	Apr	May	Jun	Jul	Aug	Sept	Oct	Nov	Dec
1956											501	1031
1957	1652	2567	3569	4564	5657	6921	8229	9676	10897	12433	14232	15804
1958	17127	18245	19192	19929								

Last engine 20553 built

Monthly First Serial Numbers—Engine Farmall 450 HC

	Jan	Feb	Mar	Apr	May	Jun	Jul	Aug	Sept	Oct	Nov	Dec
1956											600	1424
1957	1881	2860	3616			7788				12618	15087	16388
1958	17137	18862		20509								

Last engine 20509 built

Monthly First Serial Numbers—Engine International W-450

	Jan	Feb	Mar	Apr	May	Jun	Jul	Aug	Sept	Oct	Nov	Dec
1956											514	1038
1957	1667	2523	3577	4592	5680	6930	8282	9674	10900	12450	14237	15814
1958	17150	18271	19218	19937								

Last engine 20554 built

Monthly First Serial Numbers—Engine D-281 Series

	Jan	Feb	Mar	Apr	May	Jun	Jul	Aug	Sept	Oct	Nov	Dec
1956										501	503	627
1957	935	1492	2055	2600	3174	3778	4454	4811	5527	6015	6464	6931
1958	7361	7646	7966									

Production Numbers

Farmall 450	Produced for U.S.	Produced for Canada	Produced for Foreign	Total
1957	12554	15	68	12637
1958	4556	22	71	4649

Farmall 450 LPG	Produced for U.S.	Produced for Canada	Produced for Foreign	Total
	Not broken out			
1958	1019	0	0	1019

Farmall 450 HC	Produced for U.S.	Produced for Canada	Produced for Foreign	Total
1957	26	0	0	26
1958	23	1	0	24

Farmall 450 HC LPG	Produced for U.S.	Produced for Canada	Produced for Foreign	Total
	Not broken out			
1958	5	0	0	5

Farmall 450 D	Produced for U.S.	Produced for Canada	Produced for Foreign	Total
1957	4724	39	164	4927
1958	1874	30	130	2034

Farmall 450 DHC	Produced for U.S.	Produced for Canada	Produced for Foreign	Total
1957	6	0	16	22
1958	1	0	7	8

International W-450	Produced for U.S.	Produced for Canada	Produced for Foreign	Total
1957	330	39	19	388
1958	126	132	27	285

International W-450-D	Produced for U.S.	Produced for Canada	Produced for Foreign	Total
1957	255	354	94	703
1958	101	252	52	405

International W-450 LPG	Produced for U.S.	Produced for Canada	Produced for Foreign	Total
	Not Broken Out			
1958	28	0	0	28

Model 600

Yearly Serial Numbers
1956 501—1985

Monthly First Serial Numbers—Chassis All Dates 1956

	Jan	Feb	Mar	Apr	May	Jun	Jul	Aug	Sept	Oct	Nov	Dec
600D				501	656	875	1149	1411	1703	1874	1985	
600									1806	1876		

Monthly First Serial Numbers—Engine All Dates 1956

	Jan	Feb	Mar	Apr	May	Jun	Jul	Aug	Sept	Oct	Nov	Dec
600D				7507	7729	8086	8390	8699	8896	8991	9091	
600								501	502	521		

650

Yearly Serial Numbers
1956 501—687
1957 688—3451 or 688-3470 (IH sources disagree)
1958 3471—5433

Monthly First Serial Numbers—Chassis 650 D

	Jan	Feb	Mar	Apr	May	Jun	Jul	Aug	Sept	Oct	Nov	Dec
1956											501	577
1957	688	877	1067	1277	1541	1875	2223	2339	2594	2834	3062	3220
1958	3452	3672	3872	4082	4328	4580	4833	4988	5198			

Monthly First Serial Numbers—Chassis 650 Gas

	Jan	Feb	Mar	Apr	May	Jun	Jul	Aug	Sept	Oct	Nov	Dec
1957		1044	1068		1639	1883		2459	2602	2840	3068	3279
1958	3471	3776	4024	4085	4339	4591	4977	5040				

Monthly First Serial Numbers—650 LPG

	Jan	Feb	Mar	Apr	May	Jun	Jul	Aug	Sept	Oct	Nov	Dec
1957			1275	1278	1540	1876		2490	2595	2837	3070	3260
1958	3457		3996	4177	4331	4583	4832		5291			

Monthly First Serial Number—Engine International 650

	Jan	Feb	Mar	Apr	May	Jun	Jul	Aug	Sept	Oct	Nov	Dec
1956												
1957	541	542	555	580	680	704	764	769	849	934	949	990
1958	1015	1018	1023	1049	1092	1155	1170		1172	1186		

Monthly First Serial Number—Engine International 650 D

	Jan	Feb	Mar	Apr	May	Jun	Jul	Aug	Sept	Oct	Nov	Dec
1956											9091	9202
1957	9315	9522	9709	9907	10097	10419	10679	10810	10958	11123	11332	11484
1958	11659	11879	12079	12273	12471	12660	12891	13047	13230	13350		

*According to a note in the specification lists, the D-350 engine serial numbers started with 9087 for the International 650 D tractors.
Last built

Cotton Picker Tractors
Farmall –300 124 (1955), Farmall –300 Lo Drum 199 (1956), Farmall –400 127 (1955), Farmall –400 Lo Drum 135 (1956), Farmall –400 Hi-Drum 119 (1956), Farmall –400 D 2 (1955), Farmall –400 D Lo Drum 2 (1956), Farmall –400 D Hi-Drum 18 (1956), Farmall –350 Lo-Drum 126, Farmall –350 D Lo Drum24, Farmall –450 Lo Drum96, Farmall –450 D Lo Drum19, Farmall –450 Hi Drum98, Farmall –450 –D Hi Drum 27
The above tractors were built in 1957. The 1958 sheets do not mention cotton picker tractor production, but given the high levels of inventory, there may not have been 1958 production.

Serial Suffixes
IH used a series of letters located behind the serial number on both the chassis and engine serial numbers to indicate special equipment that would affect spare parts ordering and service.

Chassis Suffixes
A-distillate
B-kerosene
C-LPG
D-5,000-foot altitude
E-8,000-foot altitude
F-cotton picker mounting attachment (high drum)
G-cotton picker tractor attachment (high drum)
H-rear frame cover and shifter attachment
I-Rockford clutch
J-Rockford clutch
K-optional fourth gear
L-high-altitude cylinder head
M-low-speed attachment
N-LPG-burning attachment
O-unknown, probably an unlisted attachment

P-Independent PTO attachment without TA
Q-unknown, probably an unlisted attachment
R-TA with provision for transmission-driven PTO
S-TA with provision for 540 rpm Independent PTO
T-cotton picker mounting attachment (low drum)
U-high-altitude attachment (gasoline and LPG fuels)
V-exhaust valve rotator
W-forward and reverse drive
X-high speed low and reverse attachment
Y-Hydra Touch low and reverse attachment (with 12 gpm pump)
Z-Hydra Touch power supply (with 17 gpm pump)
AA-1,000 rpm Independent PTO drive
BB-unknown, probably an unlisted attachment
CC-third speed heavy-duty tillage gear
EE-unknown, probably an unlisted attachment

Engine Suffixes
A-distillate
B-kerosene-burning attachment
C-LPG burning attachment
D-5,000-foot altitude attachment
E-8,000-foot altitude attachment
F-unknown, probably an unlisted attachment
G-unknown, probably an unlisted attachment
HA-high-altitude heads
I-unknown, probably an unlisted attachment
J-unknown, probably an unlisted attachment
K-cast iron piston

L-high-altitude cylinder heads
M-unknown, probably an unlisted attachment
N-LPG attachment (2,500 feet and up)
O-unknown, probably an unlisted attachment
P-unknown, probably an unlisted attachment
Q-unknown, probably an unlisted attachment
R-exhaust-valve rotator attachment
S-unknown, probably an unlisted attachment
T-unknown, probably an unlisted attachment
U-high-altitude attachment
V-exhaust-valve rotator attachment

On March 21, 1957, the D and E codes were cut out from the various specification lists and replaced with the U code.

Casting Codes
Casting Codes are dates that were cast into large parts. The tag with the date code was screwed onto the patterns every day, leaving an impression in the mold, that in turn resulted in the date code showing up on the casting. The date represents the date that the part was cast, not when it was machined or assembled onto the tractor as some people believe. As parts could be cast on different days, the casting codes on a tractor will not match up on the same day, but may vary by as much as months on lower-production tractors, or on parts common to many tractors, such as belt pulley housings.

The codes are in a month, day, year format with the year being represented by a letter. The following years were used on the tractors covered in this book:
Z-1954, A-1955, B-1956, C-1957, D-1958

Appendix B
Paint and Decals

International went through several transitions in tractor appearance and styling with the tractors covered in this book. The most noticeable was the addition of emblems for make and model information, replacing decals. The sheet metal changes, especially grills and dash construction, are obvious. The change to a two tone color combination (white painted grills, and white tradename background decals) occurred in 1956. Red paint was IH-1102-B, which is barely distinguishable from IH 50 and IH 2150. Some parts began to painted separately off the tractor, then assembled onto the tractor after the main body was painted red.

As per standard IH practice, there was a listing of parts that were not to be painted during the painting process, usually due to either the vulnerability of the part to paint or the oven drying process, the need to see through the part (gauges and light lenses) or for appearance. In addition, some parts were identified as being painted by vendors or during the manufacturing process, and not to be painted during the general spray painting process.

A listing of the parts to be masked off (of left off the tractor during the painting process if masking was too difficult) includes:

1. Magneto (or distributor), spark plugs, and high tension coils
2. All electrical terminals and terminal screws
3. Chassis serial number plates and all name plates including generator and starter motor name plates
4. Glass face of indicator gauges, also lens on head and rear lamps when used
5. Fuel strainer glass bowl
6. Seat cushion
7. Back side of radiator core (apparently Louisville only had this requirement, not known if Farmall used a different assembly procedure that avoided paint the
   cores, or if they did allow overspray on the core).
8. All exposed spline shafts (see the radiator core note above)
9. Steering wheel grip
10. Resistor on underside of voltage regulator.
11. Fuse holder knob

Parts to be assembled to the tractor after painting:
Pneumatic tires and rims (Presumably including the nuts and bolts attaching the wheels to the axles)
Exhaust Muffler
Decals, name plates, and tags.

The main body of the distributor was painted red, apparently before assembly to the tractor. The cap was left black, as well as the spark plug wires. On the coil, the main body was painted red, with the caps at each end left their natural molded black.

On Farmall 200 and 230 tractors, aluminum paint was used as a primer coat for the underside of the hood (which apparently did receive primer) when the tractor was used with a cotton picker mounting attachment. The paint was from Shaffer Varnish Co, Louisville Kentucky, and was identified as their #606 Aluminum Touch-Up Paint.

In addition to paint, there were some other substances used on these tractors that may look like paint. "Clyde's All-Purpose Sealer" was used on all pipe plugs, cup plugs, and steering worm bearing retainers, as well as studs and cap screws that entered oil chambers. Welch (freeze) plugs were assembled with Tousey Varnish Company's #36 Red Gasket Sealer. Surfaces of engine, transmission, and other parts that were in contact with oil also got a coat of sealer, including the block and head. This was done mainly to seal into the casting and loose foundry sand.

Instrument panels on the 300 and 400 were painted differently than the rest of the tractor, and differently from the 350 and 450 series tractors. See the chapter text section for more information.

Appendix C
Decal Drawings

These rare drawings show decal placement as the factory intended it.

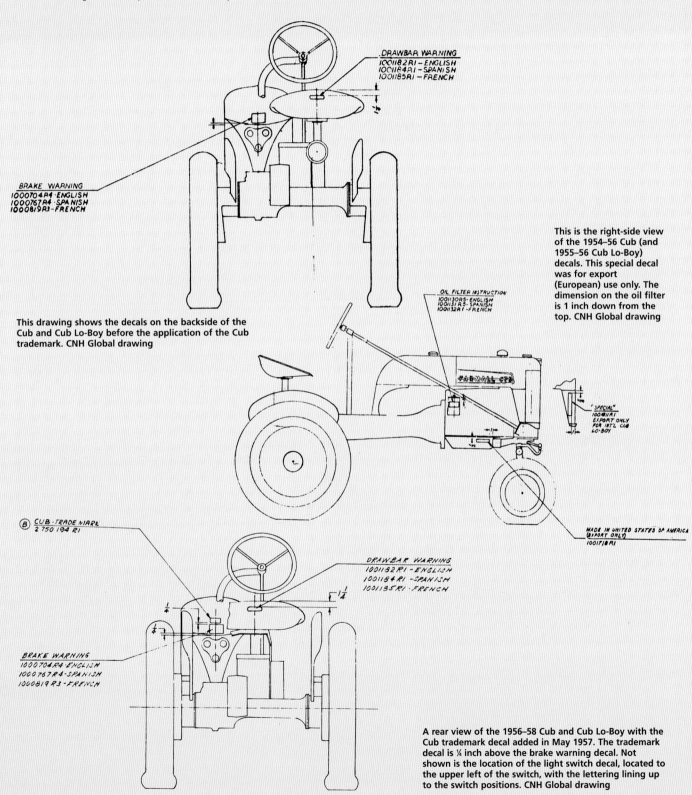

This drawing shows the decals on the backside of the Cub and Cub Lo-Boy before the application of the Cub trademark. CNH Global drawing

This is the right-side view of the 1954–56 Cub (and 1955–56 Cub Lo-Boy) decals. This special decal was for export (European) use only. The dimension on the oil filter is 1 inch down from the top. CNH Global drawing

A rear view of the 1956–58 Cub and Cub Lo-Boy with the Cub trademark decal added in May 1957. The trademark decal is ¼ inch above the brake warning decal. Not shown is the location of the light switch decal, located to the upper left of the switch, with the lettering lining up to the switch positions. CNH Global drawing

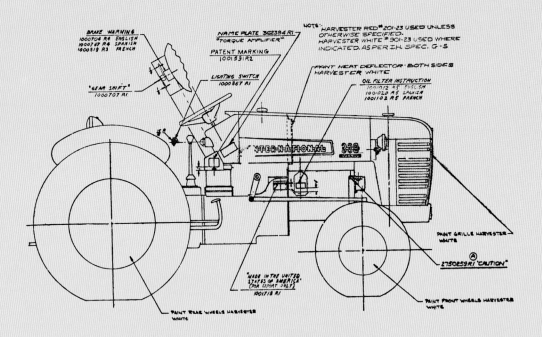

This view is of the right side of the 1956–58 Lo-Boy. Notice the white paint (Harvester White #233) on the front and rear wheels, as well as on the radiator grill. The Cub was similar, except that the wheels were not specified a different color—front wheels were red, rear wheels were red with silver rims. CNH Global drawing

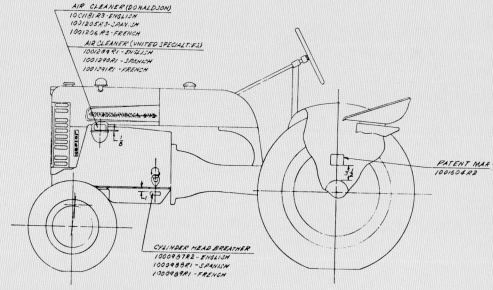

The left side of the International Cub Lo-Boy as built between 1956 and 1958. Decals from 1954 to 1956 had identical placement. CNH Global drawing

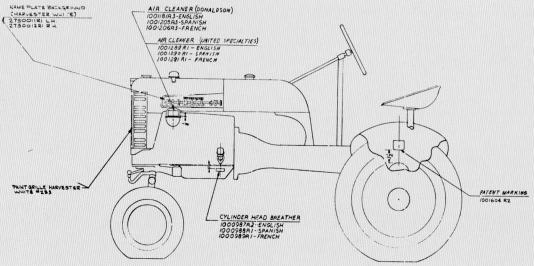

The left side of the Farmall Cub as produced from 1956 to 1958. The dimension on the two optional air cleaner decals is ⅛ inch above the lip. The dimension for the patent decal is 3½ inch above the axle, slightly different from its location on the Lo-Boy. CNH Global drawing

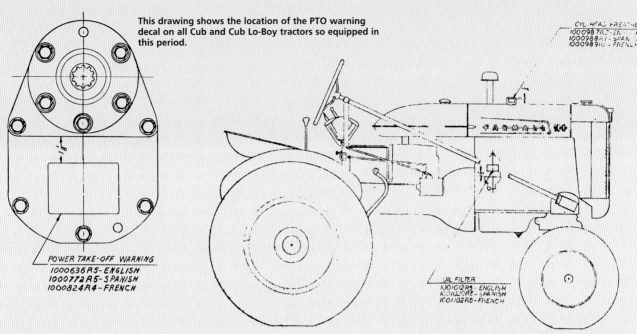

This drawing shows the location of the PTO warning decal on all Cub and Cub Lo-Boy tractors so equipped in this period.

A view of the Farmall 100's right-side decal placement drawing. Decal placements were identical to the Farmall 130, except for the 130's white-background decal on the hood, which, of course, the 100 did not have. CNH Global drawing

The top decal placements on the Farmall 130, which were identical to the placements on the Farmall 100. CNH Global drawing

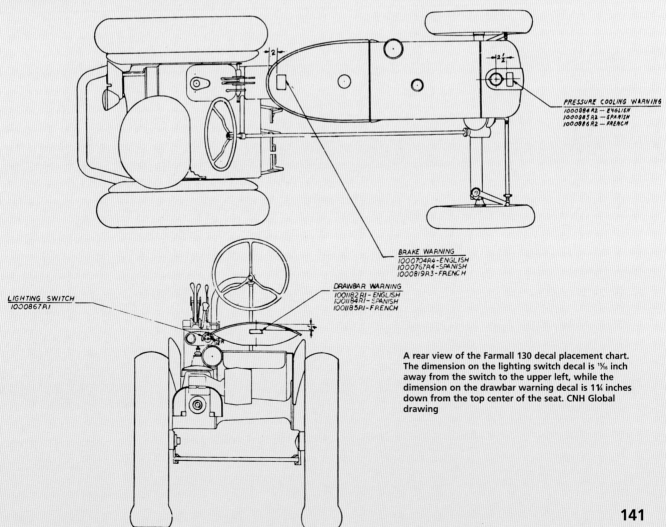

A rear view of the Farmall 130 decal placement chart. The dimension on the lighting switch decal is $15/16$ inch away from the switch to the upper left, while the dimension on the drawbar warning decal is 1¼ inches down from the top center of the seat. CNH Global drawing

141

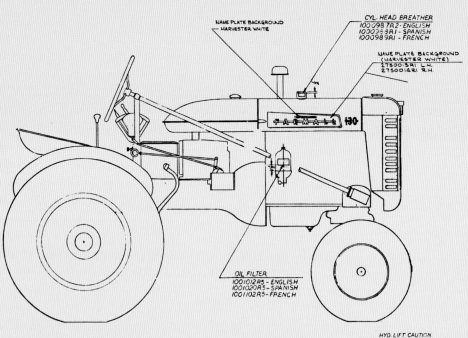

The right side of the Farmall 130 shows the white background decal. In addition, the background of the "McCormick" emblem was painted white between the letters. CNH Global drawing

The right side of the Farmall 130, which is identical to the Farmall 100 except for the white radiator grill and background decal. Of course, the International 100 and 130 were similar, having only different name emblems. CNH Global drawing

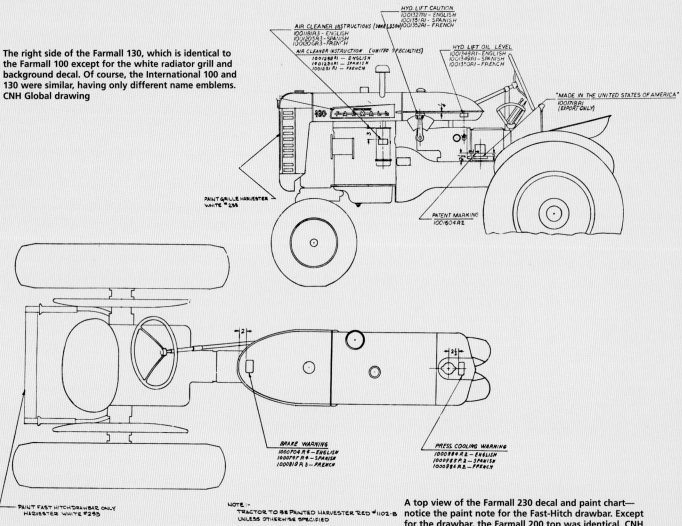

A top view of the Farmall 230 decal and paint chart—notice the paint note for the Fast-Hitch drawbar. Except for the drawbar, the Farmall 200 top was identical. CNH Global drawing

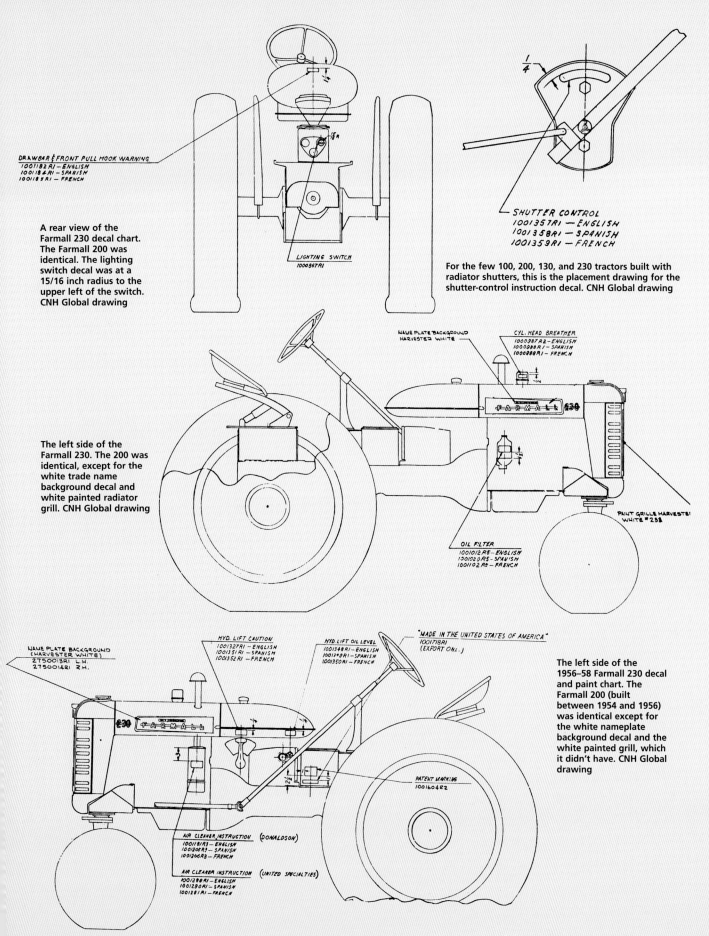

A rear view of the Farmall 230 decal chart. The Farmall 200 was identical. The lighting switch decal was at a 15/16 inch radius to the upper left of the switch. CNH Global drawing

For the few 100, 200, 130, and 230 tractors built with radiator shutters, this is the placement drawing for the shutter-control instruction decal. CNH Global drawing

The left side of the Farmall 230. The 200 was identical, except for the white trade name background decal and white painted radiator grill. CNH Global drawing

The left side of the 1956–58 Farmall 230 decal and paint chart. The Farmall 200 (built between 1954 and 1956) was identical except for the white nameplate background decal and the white painted grill, which it didn't have. CNH Global drawing

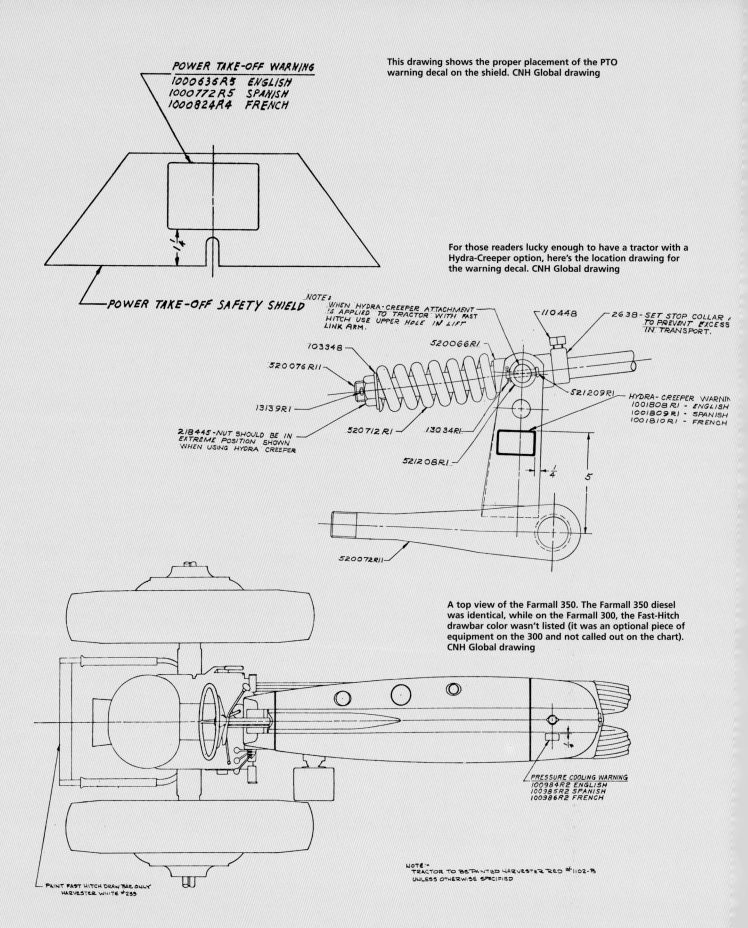

This drawing shows the proper placement of the PTO warning decal on the shield. CNH Global drawing

For those readers lucky enough to have a tractor with a Hydra-Creeper option, here's the location drawing for the warning decal. CNH Global drawing

A top view of the Farmall 350. The Farmall 350 diesel was identical, while on the Farmall 300, the Fast-Hitch drawbar color wasn't listed (it was an optional piece of equipment on the 300 and not called out on the chart). CNH Global drawing

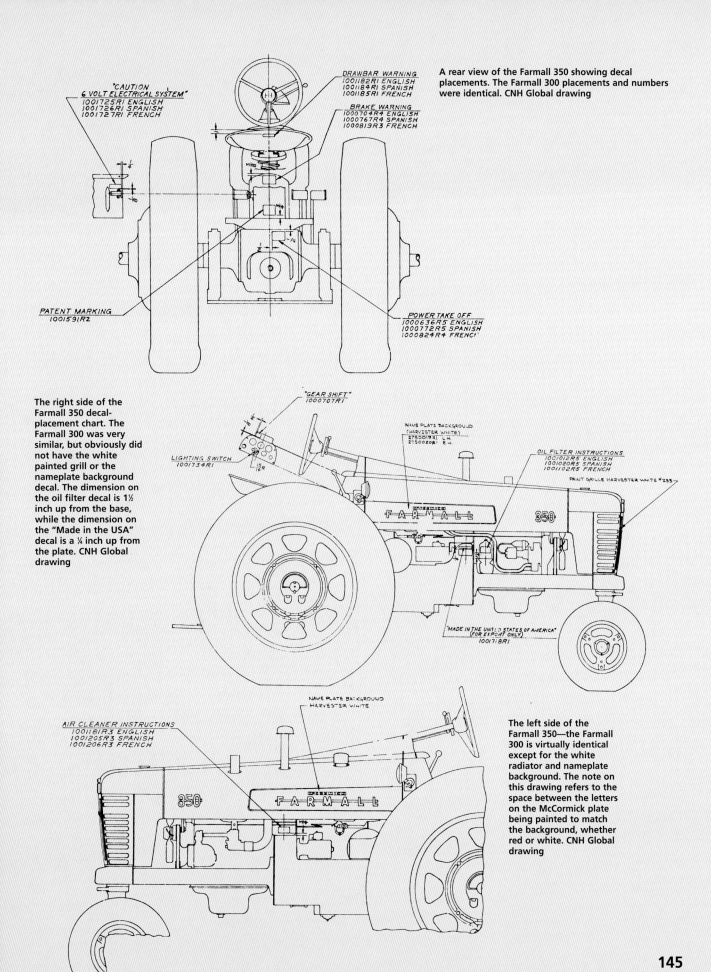

A rear view of the Farmall 350 showing decal placements. The Farmall 300 placements and numbers were identical. CNH Global drawing

The right side of the Farmall 350 decal-placement chart. The Farmall 300 was very similar, but obviously did not have the white painted grill or the nameplate background decal. The dimension on the oil filter decal is 1½ inch up from the base, while the dimension on the "Made in the USA" decal is a ¼ inch up from the plate. CNH Global drawing

The left side of the Farmall 350—the Farmall 300 is virtually identical except for the white radiator and nameplate background. The note on this drawing refers to the space between the letters on the McCormick plate being painted to match the background, whether red or white. CNH Global drawing

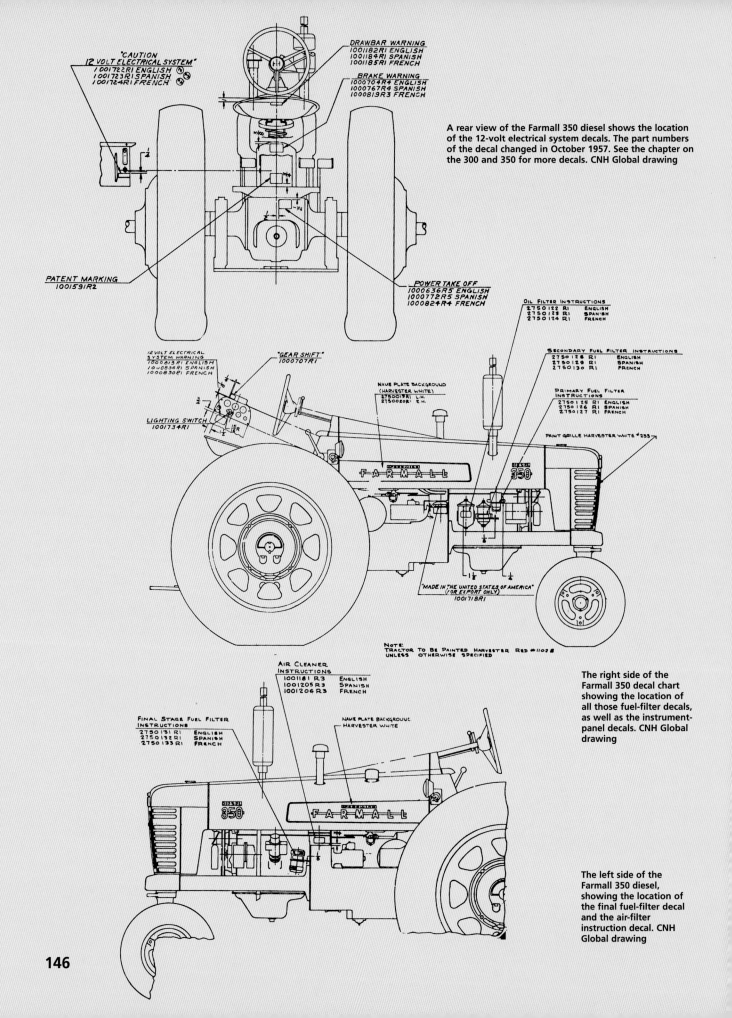

A rear view of the Farmall 350 diesel shows the location of the 12-volt electrical system decals. The part numbers of the decal changed in October 1957. See the chapter on the 300 and 350 for more decals. CNH Global drawing

The right side of the Farmall 350 decal chart showing the location of all those fuel-filter decals, as well as the instrument-panel decals. CNH Global drawing

The left side of the Farmall 350 diesel, showing the location of the final fuel-filter decal and the air-filter instruction decal. CNH Global drawing

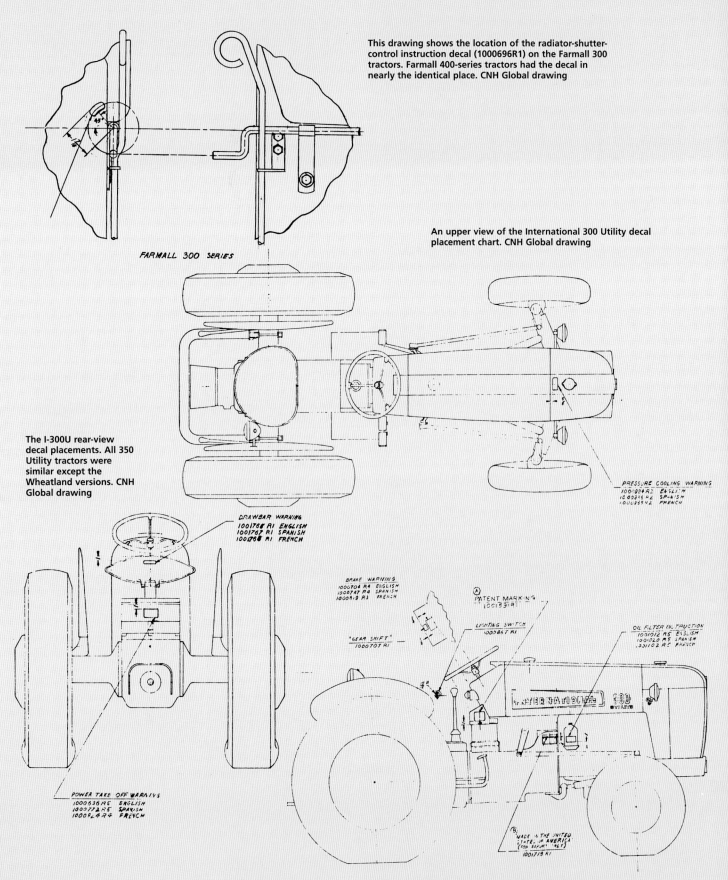

This drawing shows the location of the radiator-shutter-control instruction decal (1000696R1) on the Farmall 300 tractors. Farmall 400-series tractors had the decal in nearly the identical place. CNH Global drawing

An upper view of the International 300 Utility decal placement chart. CNH Global drawing

The I-300U rear-view decal placements. All 350 Utility tractors were similar except the Wheatland versions. CNH Global drawing

A right-side view of the International 300 Utility. The dimension on the oil-filter decal is 1 inch up from the base, as is the patent number decal. CNH Global drawing

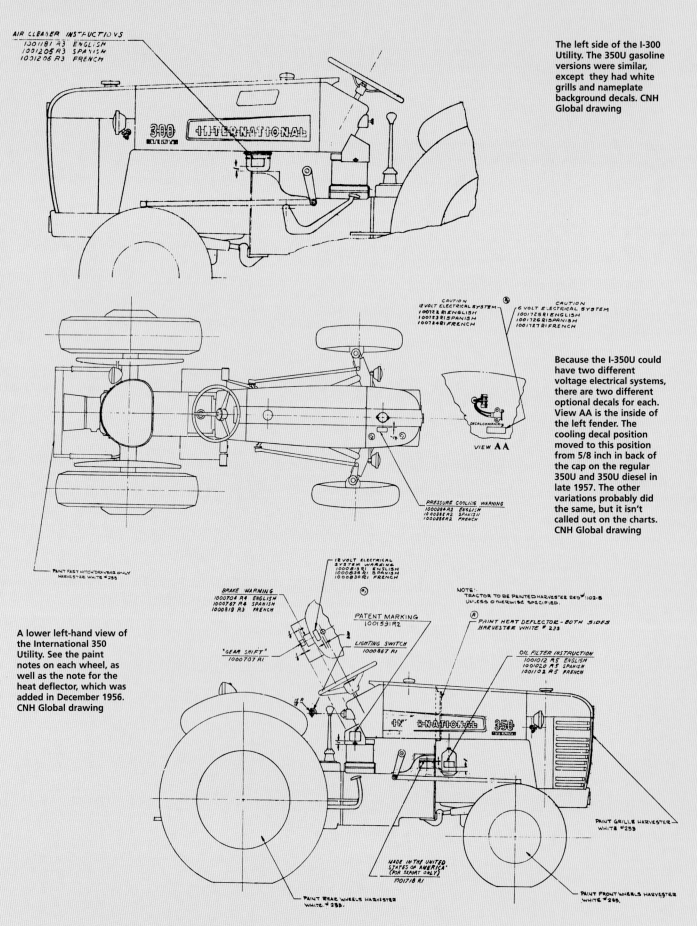

The left side of the I-300 Utility. The 350U gasoline versions were similar, except they had white grills and nameplate background decals. CNH Global drawing

Because the I-350U could have two different voltage electrical systems, there are two different optional decals for each. View AA is the inside of the left fender. The cooling decal position moved to this position from 5/8 inch in back of the cap on the regular 350U and 350U diesel in late 1957. The other variations probably did the same, but it isn't called out on the charts. CNH Global drawing

A lower left-hand view of the International 350 Utility. See the paint notes on each wheel, as well as the note for the heat deflector, which was added in December 1956. CNH Global drawing

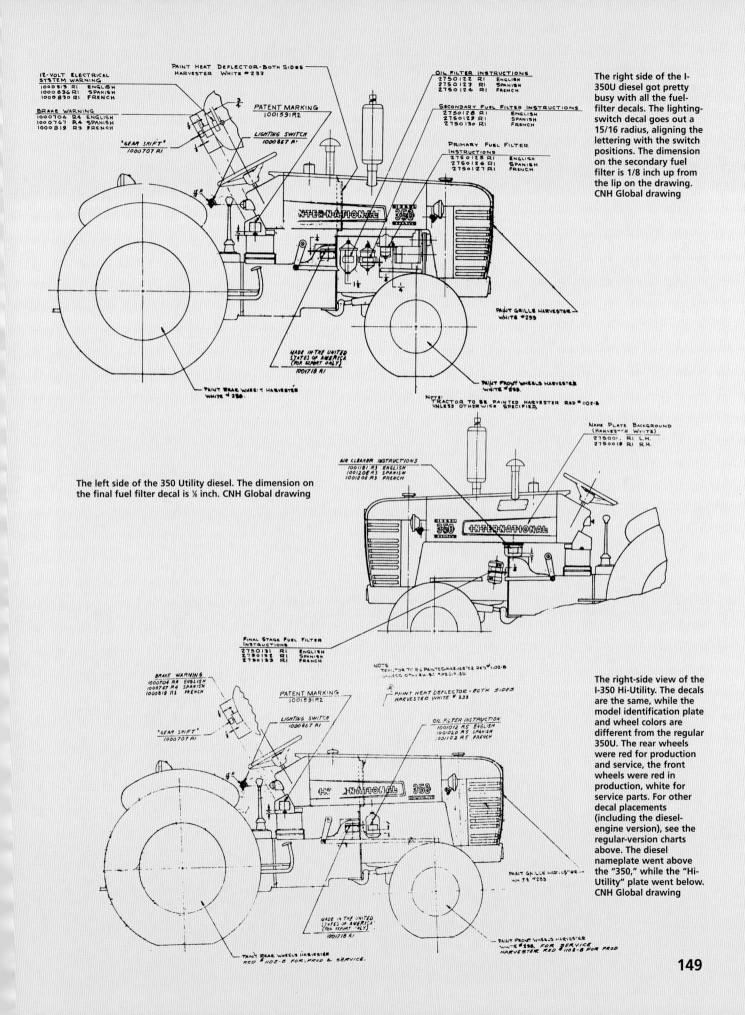

The right side of the I-350U diesel got pretty busy with all the fuel-filter decals. The lighting-switch decal goes out a 15/16 radius, aligning the lettering with the switch positions. The dimension on the secondary fuel filter is 1/8 inch up from the lip on the drawing. CNH Global drawing

The left side of the 350 Utility diesel. The dimension on the final fuel filter decal is 1/8 inch. CNH Global drawing

The right-side view of the I-350 Hi-Utility. The decals are the same, while the model identification plate and wheel colors are different from the regular 350U. The rear wheels were red for production and service, the front wheels were red in production, white for service parts. For other decal placements (including the diesel-engine version), see the regular-version charts above. The diesel nameplate went above the "350," while the "Hi-Utility" plate went below. CNH Global drawing

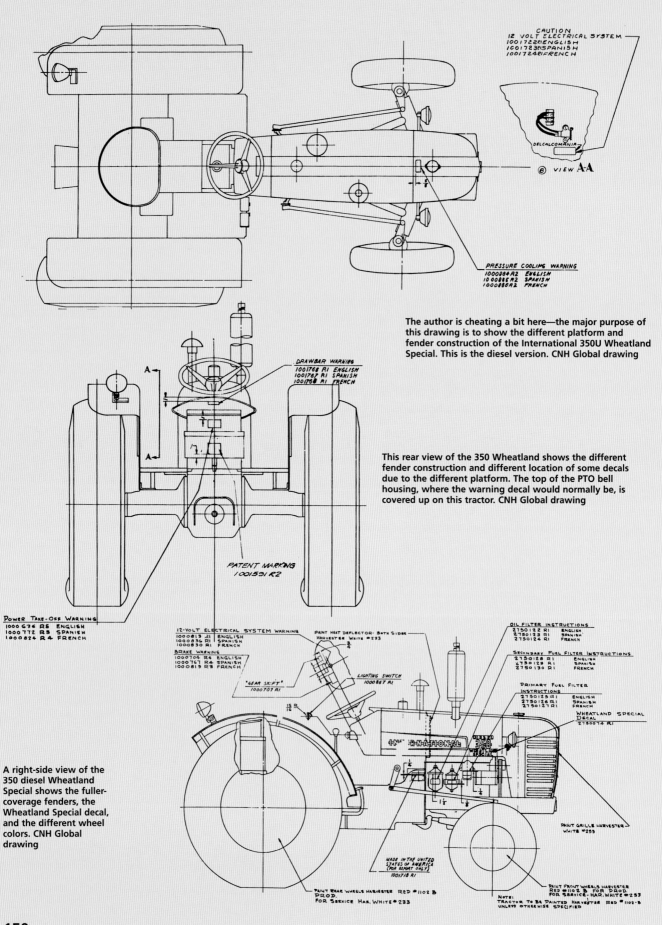

The author is cheating a bit here—the major purpose of this drawing is to show the different platform and fender construction of the International 350U Wheatland Special. This is the diesel version. CNH Global drawing

This rear view of the 350 Wheatland shows the different fender construction and different location of some decals due to the different platform. The top of the PTO bell housing, where the warning decal would normally be, is covered up on this tractor. CNH Global drawing

A right-side view of the 350 diesel Wheatland Special shows the fuller-coverage fenders, the Wheatland Special decal, and the different wheel colors. CNH Global drawing

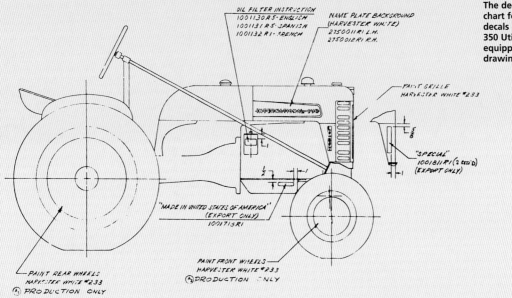

The decal-placement chart for the LPG-tank decals on the I-300 and I-350 Utility tractors so equipped. CNH Global drawing

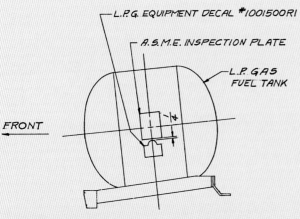

The decal location chart for the Farmall 300, 350, 400, and 450 LPG tanks. CNH Global drawing

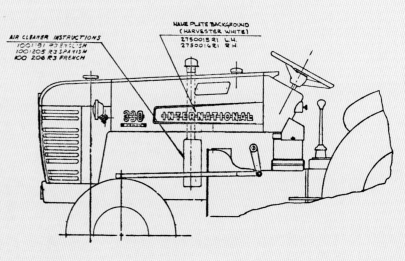

The left side of the I-330. CNH Global drawing

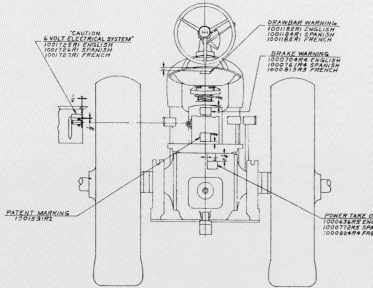

A rear view of the Farmall 450 gas-decal placement chart. The Farmall 400 gas tractor had identical placements. CNH Global drawing

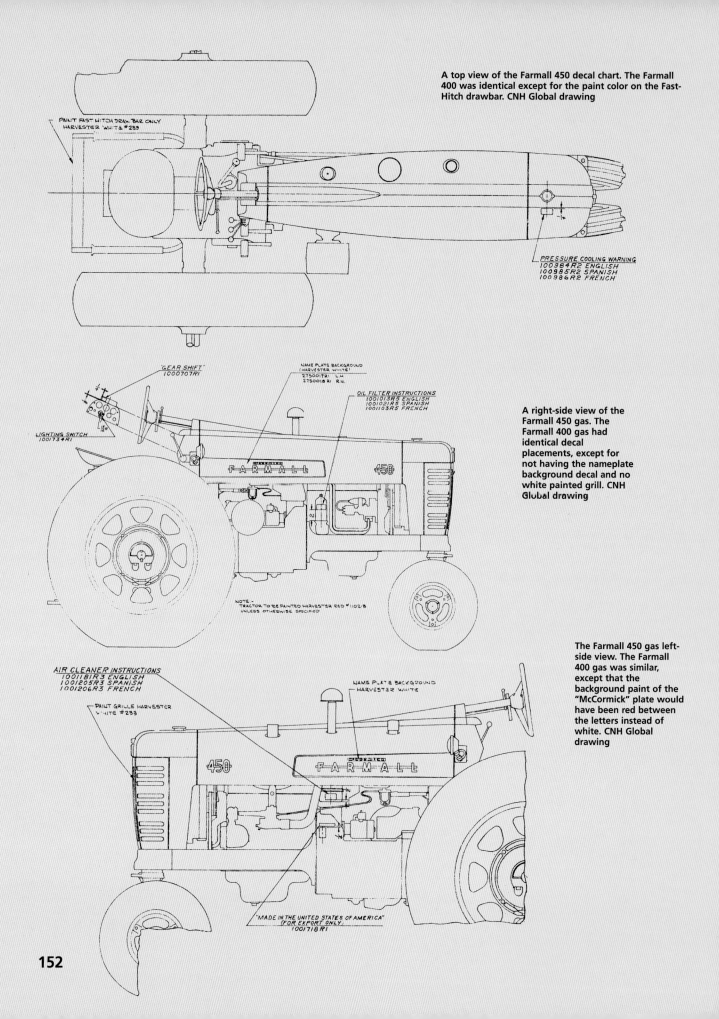

A top view of the Farmall 450 decal chart. The Farmall 400 was identical except for the paint color on the Fast-Hitch drawbar. CNH Global drawing

A right-side view of the Farmall 450 gas. The Farmall 400 gas had identical decal placements, except for not having the nameplate background decal and no white painted grill. CNH Global drawing

The Farmall 450 gas left-side view. The Farmall 400 gas was similar, except that the background paint of the "McCormick" plate would have been red between the letters instead of white. CNH Global drawing

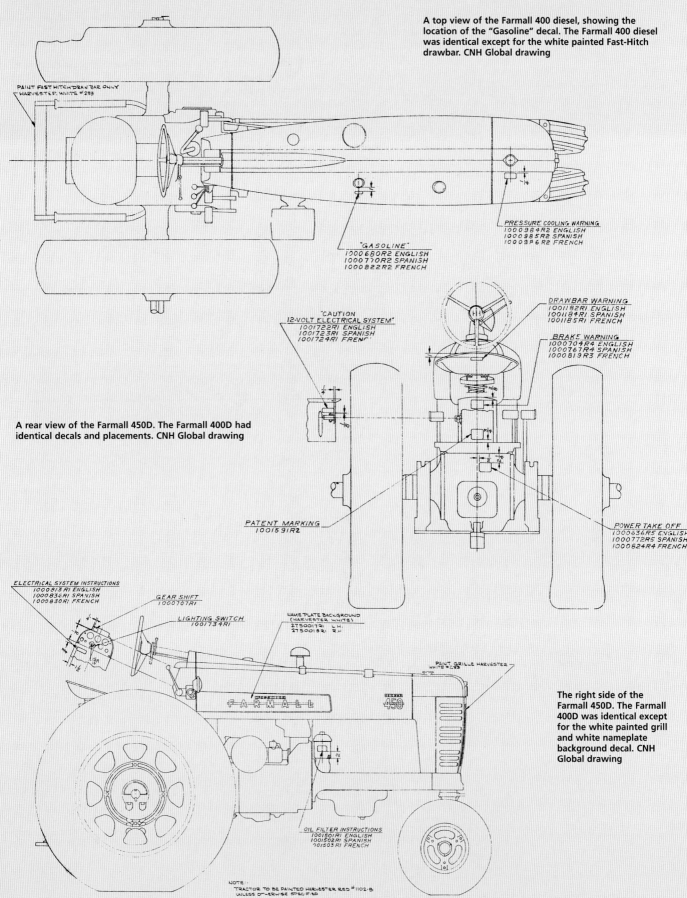

A top view of the Farmall 400 diesel, showing the location of the "Gasoline" decal. The Farmall 400 diesel was identical except for the white painted Fast-Hitch drawbar. CNH Global drawing

A rear view of the Farmall 450D. The Farmall 400D had identical decals and placements. CNH Global drawing

The right side of the Farmall 450D. The Farmall 400D was identical except for the white painted grill and white nameplate background decal. CNH Global drawing

153

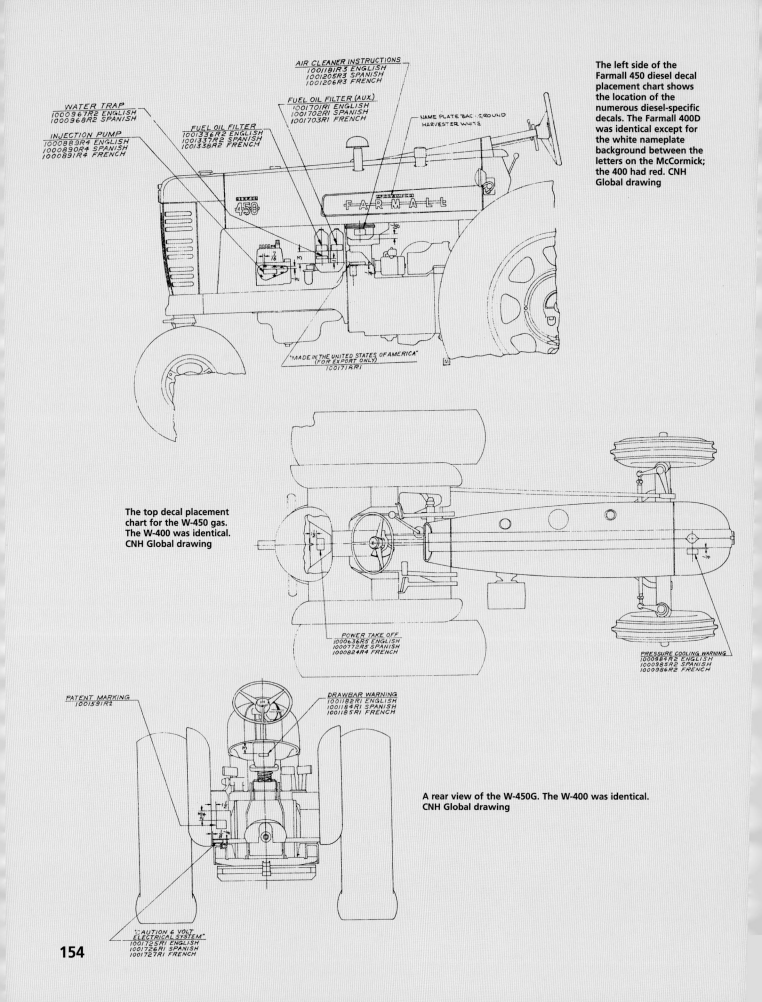

The left side of the Farmall 450 diesel decal placement chart shows the location of the numerous diesel-specific decals. The Farmall 400D was identical except for the white nameplate background between the letters on the McCormick; the 400 had red. CNH Global drawing

The top decal placement chart for the W-450 gas. The W-400 was identical. CNH Global drawing

A rear view of the W-450G. The W-400 was identical. CNH Global drawing

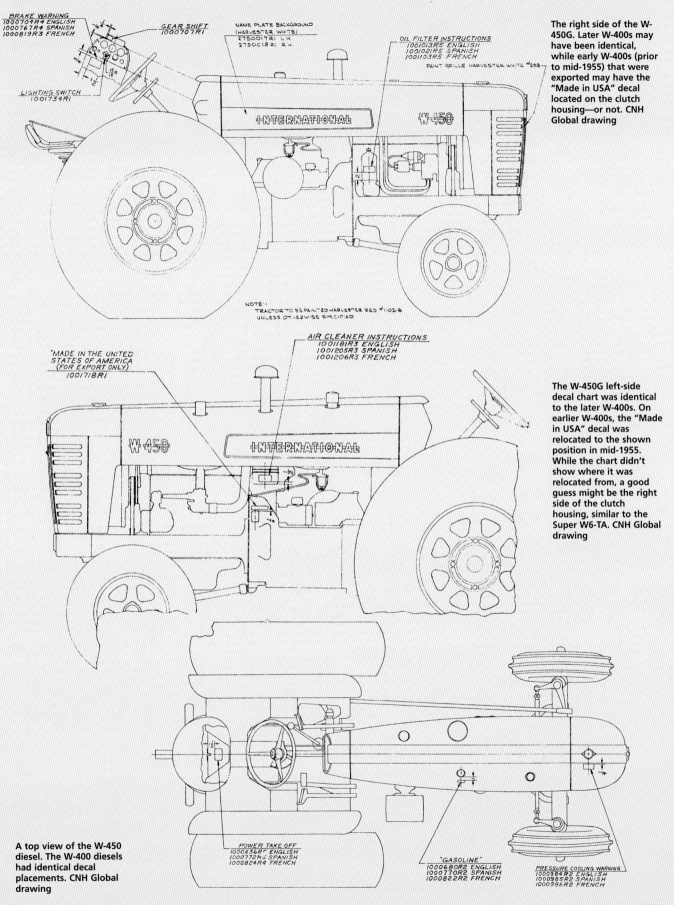

The right side of the W-450G. Later W-400s may have been identical, while early W-400s (prior to mid-1955) that were exported may have the "Made in USA" decal located on the clutch housing—or not. CNH Global drawing

The W-450G left-side decal chart was identical to the later W-400s. On earlier W-400s, the "Made in USA" decal was relocated to the shown position in mid-1955. While the chart didn't show where it was relocated from, a good guess might be the right side of the clutch housing, similar to the Super W6-TA. CNH Global drawing

A top view of the W-450 diesel. The W-400 diesels had identical decal placements. CNH Global drawing

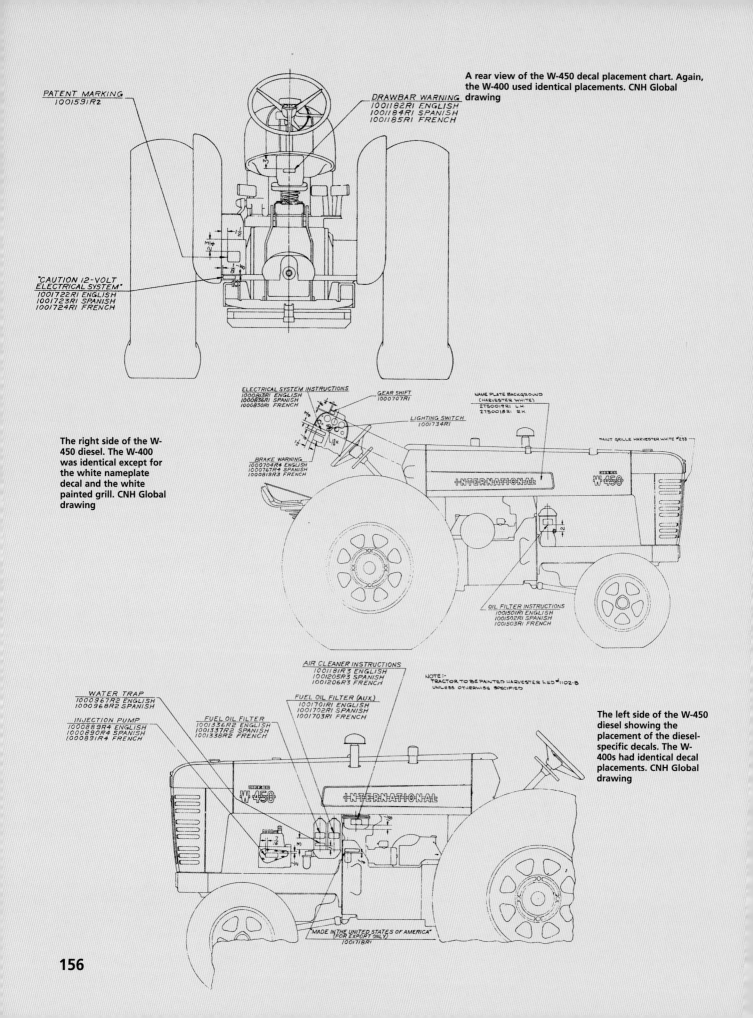

A rear view of the W-450 decal placement chart. Again, the W-400 used identical placements. CNH Global drawing

The right side of the W-450 diesel. The W-400 was identical except for the white nameplate decal and the white painted grill. CNH Global drawing

The left side of the W-450 diesel showing the placement of the diesel-specific decals. The W-400s had identical decal placements. CNH Global drawing

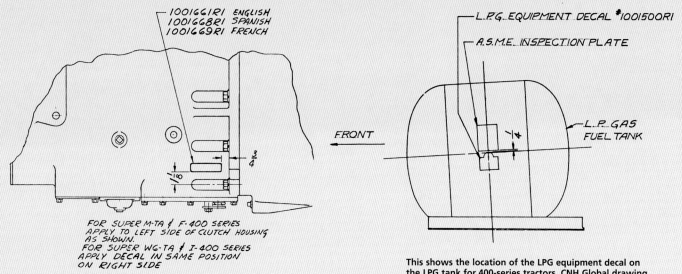

This chart shows the location of the IPTO seasonal disconnect instruction decal for the 400 series tractors so equipped. CNH Global drawing

This shows the location of the LPG equipment decal on the LPG tank for 400-series tractors. CNH Global drawing

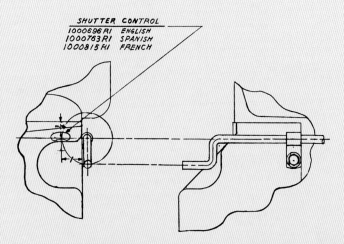

This chart shows the location of the radiator shutter control instruction decal for the International W-400 and W-450 tractors. The Farmall 400 and 450 tractors used the same decal, but the placement was a little different, similar to the Farmall 300 placement. CNH Global drawing

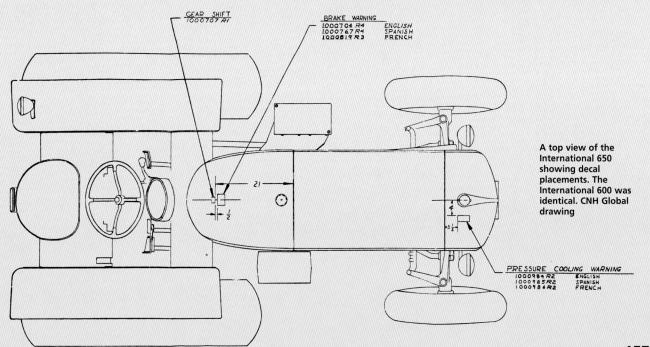

A top view of the International 650 showing decal placements. The International 600 was identical. CNH Global drawing

157

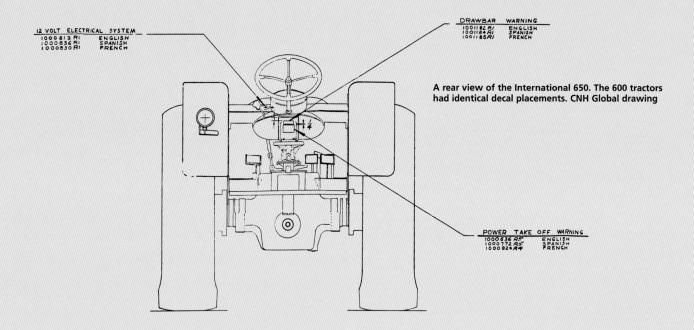

A rear view of the International 650. The 600 tractors had identical decal placements. CNH Global drawing

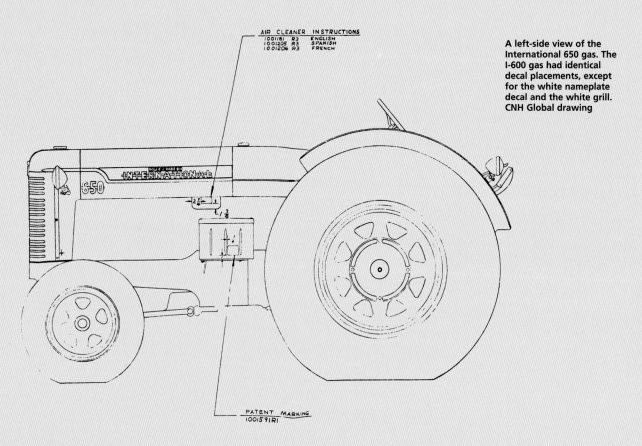

A left-side view of the International 650 gas. The I-600 gas had identical decal placements, except for the white nameplate decal and the white grill. CNH Global drawing

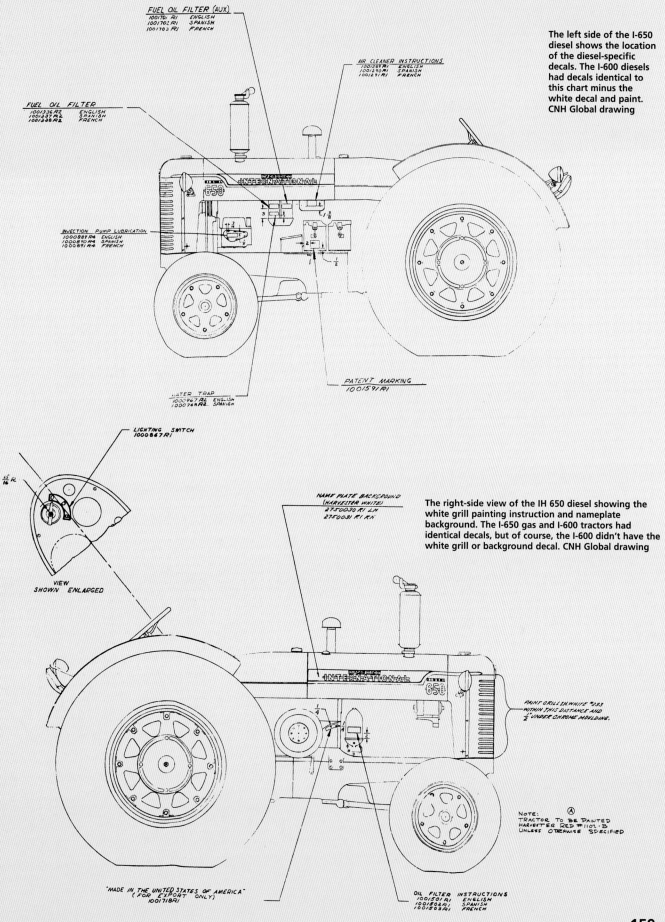

The left side of the I-650 diesel shows the location of the diesel-specific decals. The I-600 diesels had decals identical to this chart minus the white decal and paint. CNH Global drawing

The right-side view of the IH 650 diesel showing the white grill painting instruction and nameplate background. The I-650 gas and I-600 tractors had identical decals, but of course, the I-600 didn't have the white grill or background decal. CNH Global drawing

159

Index

Belt pulley, 42, 61, 100, 127
Briggs and Stratton, 117
Browning Mfg., 42, 61
Cole Hersey Co., 86, 128
Continental Motors, 48, 49
Corrie, Don, 117
Cotton pickers, 29, 30, 65, 95, 111
Cruver Mfg. Co., 56
Donaldson, 41, 42, 120, 126
Eaton Mfg. Co., 35
Electrical, 108, 109, 116, 117
Fast Hitch, 21, 22, 28, 40, 43, 47, 48, 58, 59, 62, 65, 66, 75, 87, 88, 91, 94, 105
Ford 8N, 46
Gits Molding Corp., 56
Greyer Molding Co., 56
H.A. Douglas Mfg. Co., 117
Harris, Darius, 89
Hayes Industries, 42
Hersee, Cole, 34
Hi-Utility, 66, 67
Hydra Creeper, 43
Hydra Touch, 29, 40, 41, 47, 48, 55, 58, 60, 62, 98–100
Hydraulics, 40, 41, 58, 73, 93, 94, 125
IH models,
    100 HC, 34, 39, 43–45
    100, 13, 28–43
    120A Cotton Picker, 94
    130 HC, 28, 39, 43–45
    130, 13, 28, 30–37, 39–44
    200, 13, 28, 30–33, 35–45
    230, 13, 28, 30–37, 39–45
    300 HC, 48, 53, 55, 57, 60
    300 Utility, 46, 47, 50, 52, 54, 55, 57–62
    300, 13, 46–48, 50, 52, 55–58, 60, 61, 65
    330 Utility, 72, 73, 75
    330, 74, 76
    350 D Utility, 60
    350 D, 48, 50, 55–58, 61, 66
    350 DHC, 55–57, 60
    350 HC, 60
    350 Utility, 48, 51, 57, 58, 72
    350, 13, 48–50, 53–58, 61–67
    400 D, 80, 91, 92, 94, 97, 102–104, 106, 110
    400 DHC, 80
    400 HC, 79, 91
    400, 9, 13, 47, 77, 79, 87, 91, 92, 94, 97, 98, 100–104, 106, 108, 110
    450 D, 81, 92, 94, 97, 102–104, 106
    450 HC, 83
    450, 11, 12, 63, 78–80, 84, 86, 87, 92, 94, 96, 97, 99, 102–104, 106, 108, 110
    600, 114, 115, 118–123, 129
    650 D, 128
    650, 119, 120–126, 129
    A, 29, 30
    Cub Lo-Boy, 14–20, 22– 27
    Cub, 14–26
    M, 63
    Super 9, 114
    Super A, 44
    Super A-1, 31, 34
    Super AV, 44
    Super AV-1, 44
    Super C, 28, 30, 34, 35, 44, 47
    Super Cub, 16
    Super H, 46, 47
    Super MTA, 46, 77
    Super W-4, 46
    Super W-6TA, 46
    W-400, 77, 80, 85, 88, 93, 100, 102, 104, 106, 108
    W-450, 48, 78, 82, 83, 85, 88, 93, 100, 102, 109, 112
    W-450D, 82
Indak Mfg. Co., 55
Independent power-takeoff, 61, 62
International Truck and Engine Corporations, 13
Joseph Pollack Co., 34, 86, 128
LPG tractors, 82, 83, 85, 89, 90, 109, 116, 121, 124
Mac Kenzie, 42
Maremont Auto Parts, 42
McCaffrey, John, 8
Milsco, 39
Monroe, 39
One-point Fast Hitch, 24
Orchard tractor, 66
Pesco Products Co., 40
Power steering, 114, 115
Power steering, 127
Power takeoff (PTO), 25, 34, 42, 43, 46, 56, 60, 61, 66, 75, 91, 105–108, 127, 128
Schrader and Engineair, 43
Stant, AC Spark Plug Co., 35, 54, 86
Thompson Products Co., 40
Touch Control, 24, 28, 34, 40
Tousey Varnish Company, 54
United Specialties, 41, 42, 117, 120, 126
W.H. Salisbury & Co., 35
Wisnefske, Rick, 64, 74
Wittek, 35